Springer Theses

Recognizing Outstanding Ph.D. Research

For further volumes:
http://www.springer.com/series/8790

Aims and Scope

The series "Springer Theses" brings together a selection of the very best Ph.D. theses from around the world and across the physical sciences. Nominated and endorsed by two recognized specialists, each published volume has been selected for its scientific excellence and the high impact of its contents for the pertinent field of research. For greater accessibility to non-specialists, the published versions include an extended introduction, as well as a foreword by the student's supervisor explaining the special relevance of the work for the field. As a whole, the series will provide a valuable resource both for newcomers to the research fields described, and for other scientists seeking detailed background information on special questions. Finally, it provides an accredited documentation of the valuable contributions made by today's younger generation of scientists.

Theses are accepted into the series by invited nomination only and must fulfill all of the following criteria

- They must be written in good English.
- The topic should fall within the confines of Chemistry, Physics, Earth Sciences, Engineering and related interdisciplinary fields such as Materials, Nanoscience, Chemical Engineering, Complex Systems and Biophysics.
- The work reported in the thesis must represent a significant scientific advance.
- If the thesis includes previously published material, permission to reproduce this must be gained from the respective copyright holder.
- They must have been examined and passed during the 12 months prior to nomination.
- Each thesis should include a foreword by the supervisor outlining the significance of its content.
- The theses should have a clearly defined structure including an introduction accessible to scientists not expert in that particular field.

Elizabeth Song Lockard

Human Migration to Space

Alternative Technological Approaches for Long-Term Adaptation to Extraterrestrial Environments

Springer

Elizabeth Song Lockard
Environmental+Interior Design
Chaminade University
Honolulu, HI, USA

ISSN 2190-5053 ISSN 2190-5061 (electronic)
ISBN 978-3-319-05929-7 ISBN 978-3-319-05930-3 (eBook)
DOI 10.1007/978-3-319-05930-3
Springer Cham Heidelberg New York Dordrecht London

Library of Congress Control Number: 2014936838

Printed on acid-free paper

Springer is part of Springer Science+Business Media (www.springer.com)

Supervisor's Foreword

I have taught graduate level courses and chaired or served as a member of scores of PhD dissertations at the University of Hawaii at Manoa and elsewhere for more than 40 years. I have also been involved in delivering core lectures and departmental workshops in "space humanities" at the International Space University, headquartered in Strasbourg, France. The dissertation by Liza Lockard is one of the most broadly conceived, diligently researched, and elegantly argued space-oriented dissertations I have ever had the privilege of chairing. It deserves to be published by Springer in this series so that it can be widely shared and admired.

Before earning a PhD in political science, Dr. Lockard was educated to be an architect, and she both teaches and engages in architectural design professionally. She thus brings an architect's sense of design, interconnectedness, balance, and dynamics to all her work, especially to this dissertation concerning the past and futures of human space flight and what it might mean for the futures of humanity, which is not only very broad in scope and foresight, but also deep in sophistication and detail. And yet it is written in an open and easily accessible style.

Most of the vast literature about human space orientation is either imminently practical or objectively scientific. Some of the literature is expansively philosophical and aspirational, often suggesting that space exploration and settlement is a necessary next step in human evolution; indeed, that the very purpose of the evolution of life itself on Earth was to evolve to the point where intelligence was able and willing move off of Earth into the inner solar system, and then eventually to other planets and in the very long run, to spread throughout the universe.

The monograph by Prof. Lockard in a sense embraces aspects of all of these views, from the most practical and "down to earth," if you will, to the most speculative, spectacular, and inspiring. Moreover, she is quite unusual in including a serious consideration of the ethical and political implications of space exploration, both immediately for Earth itself as well as for the long range futures of intelligent life in the cosmos.

Honolulu, HI — Jim Dator

To my father, Arthur Hunt Lockard, who encouraged me to always follow my heart, despite the impractical paths it frequently took me down; and to Ginger, Laka, and Scout, who stayed patiently by my side throughout my journey, offering equal amounts of moral support and necessary distraction.

They have provided me with the greatest examples of empathy, intelligence, humor, and love.

Abstract

As humans embark upon the next phase of Space exploration—establishing human outposts in low-Earth orbit, on the Moon, and on Mars—the scope of human factors must expand beyond the meager requirements for short-term missions to Space to include issues of comfort and well-being necessary for long-term durations. However, to habitate—to dwell in a place—implies more than creature comforts in order to adapt. Human factors research must also include a phenomenological perspective—an understanding of how we experience the places we live in—in order for a community to be robust and to thrive.

The first phase of migration will be an especially tenuous one requiring intensive technological intervention. The modes by which those technologies are implemented will have significant bearing on the process of human adaptation: the nature of the mediation can be either one of domination, subordination, avoidance, or integration. Ultimately, adaptation is best ensured if symbiotic processes of negotiation and cooperation between subject and environment are espoused over acts of conquest or acquiescence.

The adaptive mechanisms we choose to develop and employ will have wider implications for long-range human evolution. The transformations we will undergo will be influenced by both the *initial decision* to migrate to Space (technological), as well as the *actual conditions* of Space (environmental). Migration to extraterrestrial environments will be unequivocally the most profound catalyst for evolution in the history of humankind—not only for the human species itself but also for the new environments we will eventually inhabit. At the same time, we also find ourselves—via a new generation of bio-, nano-, and digital technologies—in the position to consciously and willfully direct our own evolution. Technology has always been transformative, but in the not-so-distant future, we will soon possess the capacity to radically re-invent ourselves in almost any way conceivable. The discourse on human evolution in Space must be situated in the confluence of these two variables.

List of Abbreviations

AI	Artificial Intelligence
AL	Artificial Life
ASI	Italian Space Agency
CAVE	Cave Automatic Virtual Environment
CELSS	Closed Ecological Life Support Systems
CNES	French National Center of Space Research
CNSA	China National Space Administration
CSA	Canada Space Agency
DARPA	Defense Advanced Research Projects Agency
ERV	Earth Return Vehicle
ESA	European Space Agency
EVA	Extra-Vehicular Activity
FMARS	Flashline Mars Arctic Research Station
GCR	Galactic Cosmic Rays
GRIN	Genetic, Robotic, Informational, and Nano-technologies
HCC	Human-Centered Computing
HDU	Habitat Demonstration Unit
HI-SEAS	Hawaii Space Exploration Analogs & Simulations
HMD	Head-Mounted Display
ICE	Isolated and Confined Environments
ISRO	India Space Research Organization
ISRU	In-Situ Resource Utilization
ISA	Israeli Space Agency
ISA	Iranian Space Agency
ISS	International Space Station
JAXA	Japan Aerospace Exploration Agency
KARI	Korea Aerospace Research Institute
LAPAN	Indonesia National Institute of Aeronautics and Space
LBNP	Lower Body Negative Pressure
LEO	Low-Earth Orbit

MDRS	Mars Desert Research Station
MER	Mars Exploration Rover
MSRM	Mars Sample Return Mission
NASA	National Aeronautics and Space Administration
NSAU	National Space Agency of Ukraine
RKA	Russian Federal Space Agency (also known as Roscosmos)
SPE	Solar Particle Event
STEM	Science, Technology, Engineering, and Mathematics
TVSS	Tactile Visual Sensory Stimulation
UKSA	United Kingdom Space Agency
VE	Virtual Environment
VR	Virtual Reality
VT	Virtual Technology

Contents

List of Figures

Chapter 1
Introduction

1.1 Caveat

The language of architecture and the language of astronomy both include the word 'space' in their respective lexicons; there is some overlap in their meaning, but also some critical subtle differences. Since this dissertation spans both those fields, I have elected to make a distinction between the two by using 'space' to represent the general condition of emptiness, absence, or void; and 'Space' to represent what we commonly refer to as outer space. The context in which the terms are used doesn't always make clear which one I am referring to, and I wanted to preserve the nuances of each. I've purposely avoided using the phrase 'outer space' altogether, as my thesis argues that this term of externality is no longer appropriate in the discourse. In the same spirit, I have also capitalized a number of other terms for clarity, treating them as names rather than as simple nouns: Nature/nature, Earth/earth, Moon/moon, Sun/sun, and Universe.

1.2 Premise

> The Earth is the cradle of humanity, but [hu]mankind cannot stay in the cradle forever.
> —Konstantin Tsiolkovsky, father of modern astronautics

> A human mission to Mars is today the ultimate destination in our solar system for humanity, and it is a priority for NASA. Our entire exploration program is aligned to support this goal.—NASA Administrator Charles Bolden

We are on the threshold of enormous change as humans embark on a new generation of Space exploration. Within the next two decades, NASA, ESA, and other international Space agencies are planning to establish the first human outposts on Mars and the Moon. The initial crews for these outposts will be small—somewhere between three and eight—and their length of stay may last only several

E.S. Lockard, *Human Migration to Space*, Springer Theses,
DOI 10.1007/978-3-319-05930-3_1, © Springer International Publishing Switzerland 2014

months or up to several years before eventually returning to Earth. But with time, as the outposts develop into settlements, and settlements later become communities, the durations of inhabitation will extend to lifetimes, and even generations.

The prospect of our species inhabiting planets other than Earth in the near future raises several questions which demand thoughtful consideration: How can we not only thrive in the short-term (single generation), but also adapt and endure in the long-term (multiple generations)? What kind of relationship to the Space environment will best serve that objective? What will non-Earthling humans be like—and could they still be called 'human' at all?

Human migration to Space will be the catalyst for the most profound transformation in the evolutionary history of humankind. And yet while we find ourselves on the cusp of such a monumental event, there has been almost no dialogue about what the impact to humankind might be—or what it ought to be. In a foreseeable future, as pioneers become residents, residents become citizens, and citizens become natives, our descendants will be born and raised somewhere other than on Earth. Their ancestral lineage will be human, but they themselves will not be human in the sense we think of being human today.

Science fiction invents numerous imaginative, sometimes bizarre, alien personae encountered in Space travels, but human beings in contrast are almost always depicted virtually identical to how we are now. Ironically, the fictional *human* characters are the least believable of them all. It is impossible to imagine that after centuries have passed, with technologies that would make our present-day inventions seem downright primitive, and the inhabitation of distant extraterrestrial venues so different from Earth, Space-humans would still resemble the Earth-humans of today. Even though *Homo sapiens* will be in fact the 'aliens' in Space, we still conceive our human characters to look like us because we need to identify with them. The notion that humans might change to such a degree that they no longer bear any resemblance to what we are today is deeply unsettling on a visceral level. There is nothing more threatening than questioning one's own identity. And yet there can be no doubt that we are beginning an era—what Dawkins refers to as the "Space travel threshold" in human evolution (Gardner 2003, p. 115)—that will effect greater transformation than anything during the entire history of our species.

1.3 Scope and Goals

The purpose of this dissertation is twofold: (1) to analyze what conditions are conducive to long-term adaptation in Space, and then evaluate various technological strategies that will be most likely to realize this goal; and (2) to assess the impact of those strategies on the evolutionary future of humankind.

Because I advocate for one specific approach over other alternatives, there is a prescriptive aspect to the dissertation. However, the normative component is limited to what kind of relationships we should cultivate with the extraterrestrial environment, focusing on technological processes; it is not concerned with what kind of entities we should become. Therefore the chapter on the implications for human evolution is exclusively of an exploratory nature. Rather than conjuring visions of what we think humans of the future will or ought to be like, adopting a bottom-up approach—in which the means are given but not the ends—allows the opportunities for emergence and transcendence to flourish unencumbered by a bias towards a particular outcome.[1]

The dissertation does not attempt to define a future being or in any way prescribe the products of evolutionary processes; by invoking the term *Homo caelestis*, my aim is only to refer to the *potentiation* of a future being, but without giving it any normative or descriptive specificity. Rather, the intent of the dissertation is to offer strategies for adaptation within a proposed theoretical framework that I contend will best promote robustness and longevity. Moreover, attempts to predict what humans of the future will be or should be inevitably result in one-dimensional caricatures that do not capture the complexities which arise from a deeper understanding of conditions that cannot yet be known. We cannot even conceive of what human immigrants to Space will become, much less describe them in any meaningful way; we can only decide *how* they should best conduct their lives on their path to becoming, and not *what* they should become.

Despite the extraordinary possibilities for human exploration in Space, there are many who for a variety of reasons vehemently believe that we should not extend our presence into the solar system. Some of the more common objections are:

- We should first concentrate our efforts on fixing the problems here on Earth, such as environmental pollution, poverty, social inequity, climate instability, etc.
- We will only transfer our destructive behaviors to Space.
- Space exploration is too costly and offers no measurable gains to humankind.
- Space travel is too risky for humans and would be best explored by robots instead.
- The conditions of Earth gave rise to human life and therefore we belong here—not on other planets where conditions are hostile to life.

[1] This approach subscribes to the methods used in artificial life simulations. In an attempt to create complex computer systems that simulated those of living systems, AL researchers began with programs of simple algorithms, and by applying them in a recursive fashion, could generate systems of complexity in much greater magnitudes of order than could possibly be quantified or predicted given the original set of rules. This method is the opposite of the approach employed in artificial intelligence research, in which human intelligence was the benchmark by which the engineered systems were assessed. The fundamental difference between artificial life and artificial intelligence resides in the notion of emergence: whereas AI—in a top-down approach—set human intelligence and consciousness as the bar by which machines were to be measured, "the goal of AL is to evolve intelligence within the machine through pathways found by the 'creatures' themselves" (Hayles 1999, p. 239). AL took the bottom-up stance that consciousness is not at the core of the human being, but rather an emergent property derived from incremental adaptive steps that the machine then learns from and builds upon. A bottom-up approach does not limit the potential of what the human being can evolve into, as the top-down approach does.

Beyond these objections, there are present circumstances that might preempt, or at least indefinitely postpone, humans landing on another planet. Recently the world has endured staggering economic crises which are no longer limited in scope to a single country or nation, as they had been in the past when economies were more isolated and not as interdependent as they are today. Recessions tend to be of a larger magnitude and have greater global repercussions. A costly and complex endeavor—one that is seen as more discretionary than the military—might easily be shelved in the event of another long-lasting economic downturn. Priorities may also shift to address more pressing environmental problems, such as increasingly severe climate instabilities and sea level rise precipitated by global warming. Another possible obstacle is the Earth's dwindling resource supply, which may render the luxury of further Space exploration prohibitive, unless alternative energy sources are developed. War or political strife could also end up taking attention away from Space endeavors—or it may further military interest and involvement in Space exploration for strategic purposes. In either case, the result would likely be a shift away from human habitation.

In addition to the possible obstacles on Earth, there are other factors that may bring human Space exploration to a halt: a large magnitude Space-related catastrophe involving human casualties could very well undermine the confidence and resolve in the Space exploration program. Or if first genesis life is discovered on Mars, we may choose to protect it by ceasing all efforts to inhabit the planet.

However, there are also circumstances that are giving more momentum to the possibility of humans living elsewhere than on Earth. At present we are enjoying a resurging interest in Space activities, paralleling STEM initiatives to promote science, technology, engineering, and mathematics curricula in academia. This recent upwelling of enthusiasm, particularly in the idea of humans visiting Mars, is due in large part to the highly publicized events of the last several years, including the discovery of several habitable exoplanets in other solar systems and the landing of the Mars Curiosity Rover, which has provided the public far more information and images of the planet than previous rovers have. The Space industry in its early days was an elitist endeavor, reserved for wealthy nations and a highly-trained, military-bred astronaut corp. Since then, the industry has opened up immensely, allowing participation of laypeople (albeit only those with considerable financial resources) and partnering with the private sector, both of which in turn generates greater zeal.

The question of whether humans will or should venture into Space does not fall within the scope of this dissertation. I begin with the assumption that not only is it inevitable that we will inhabit places beyond Earth, but that it is imperative that we do. First, all life is motile; part of life's evolutionary programming is to seek out new ecological niches. Migration to new habitats gives species a better chance of survival. Second, humans are natural explorers—which is in part why we have managed to persist as a species. To choose not to explore would be antithetical to our nature. And third, we are arguably more likely to solve problems that affect Earth if we have the opportunity to see them from a distant perspective. However, to the extent that some of the objections mentioned previously are directly related to my arguments about technology and evolution, these points will be addressed in Chapter 6.

Ultimately, the more appropriate question this dissertation addresses is not *whether* we will or should explore Space, but *how* we should explore it.

1.4 Novelty and Timeliness

1.4.1 Novelty

Space architecture is a relatively new, but quickly growing field. Until recently the design of spacecraft was the exclusive domain of the rocket scientist and technical engineer corps. The introduction of an architectural perspective was intended to incorporate human factors criteria into the design of occupied spacecraft. However, little if any, of the existing literature on Space architecture addresses the design of the habitat from the human point of view. While it exhaustively details the engineering solutions associated with the physiological challenges of spaceflight, it overlooks, for the most part, a comprehensive social and psychological perspective. Though there has indeed been extensive research done on the subject of psychological and social needs of astronauts spending long durations in Space, the subject has been limited predominantly to intra-crew dynamics and crew composition. How the physical structure of the Space habitat itself or use of other environment-mediating technologies might influence the psychological and social states of the crew (which falls within the realm of environmental psychology) has not been given much consideration.

The more recent research on habitability addresses some of these oversights, but still tends to focus on optimizing creature comforts as a means to offset the 'hardships' of living in Space. Design is above all else a problem-solving process that optimizes solutions within a given set of constraints. But an essential experiential aspect is missing: How do humans perceive Space and how do those perceptions affect the ways in which we both design and experience our habitats? How should the constructed habitat mediate the relationship to the planetary environment? How can we enable a richer experience of the Martian or lunar landscape? Though these questions fall outside NASA's narrowly defined scope of habitability, understanding these issues is crucial not only to the survival of the human crew, but also to the futures of humankind in Space. It would be imprudent to set goals for inhabitation of extraterrestrial sites without a vision for long-term adaptation and an assessment of how the strategies we employ will bear on our evolution.

This dissertation will pick up where the existing habitability research leaves off. First, in identifying the deficiencies in the current approach; second, in making an argument for incorporating an adaptational perspective—the criteria of which are qualitatively different from those of habitability—as a means to redress those deficiencies; and third, in examining various technologies that will be applicable towards that end and how the artifices we choose—but more importantly how we implement them—will impact human evolution.

1.4.2 Timeliness

After the dramatic Apollo missions of the late 1960s and early 1970s, public interest in Space exploration diminished, and although the Space stations that followed in their wake have seen their share of international visitors, not a single human being has set foot on the Moon since. The tragic fates of the Challenger (1986) and Columbia (2003), in which 14 astronauts altogether were killed, were huge setbacks for the Space program and caused the public to question both the investment and the benefits of Space exploration.

In 2004, in an attempt to revitalize the waning Space program, President Bush announced the Vision for Space Exploration. The Vision put forth the following objectives:

- *Implement* sustained and affordable human/robotic program to explore the solar system and beyond.
- *Extend* human presence across the solar system, starting with a human return to the Moon by the year 2020, in preparation for human exploration of Mars and other destinations.
- *Develop* the innovative technologies, knowledge, and infrastructures both to explore and to support decisions about the destinations for human exploration.
- *Promote* international and commercial participation in exploration to further US scientific, security, and economic interests.

NASA's Constellation program was formed as part of this new vision. It set its sights on establishing human outposts on the Moon in the 2020s and later on Mars by the mid to late 2030s. Similar programs, like ESA's Aurora program were also created with the same agenda.

Sending humans to live on Mars or the Moon is not so much a question of technological feasibility, as one of political and public will. Funding for such an ambitious undertaking, as well as the necessary technological expertise and development, will very likely come from the private sector, which will either sponsor these ventures independently or in conjunction with national Space programs. Twenty years may seem a long way out, but by Space industry standards where it takes several years just to develop a spacecraft prototype, it is in fact a very short period of time. Now is the time to expand research parameters.

Numerous events of the past decade have also rekindled the public's interest in the Space program. The discovery of Earth-like exoplanets located in the habitable zone of their respective solar systems offer growing promise that life can at least be sustained (if not already exist) on planets other than Earth. The abundance of water-ice coupled with new knowledge of the viability of extremophiles support the likelihood of past life on Mars and the increasing possibility that life may even exist there now. The debate over whether Voyager has officially exited the solar system (which was prematurely announced in March 2013) indicates that very soon it will become the very first human-made Spacecraft to cross over into interstellar Space. And the successful landing of the Mars Curiosity Rover in August 2012 begins a new period of in-depth reconnaissance of the planet, including conditions of habitability. These recent discoveries and achievements have bolstered the renewed commitment to further exploration of Space, and the prospect of establishing a human presence on other planets can be expected to generate even greater interest.

1.5 Summary of Research

The dissertation is a multi-disciplinary project that envelopes themes in architecture, interior design, environmental psychology, geography, anthropology, astrobiology, and aesthetics; as well as theories of environment, cosmology, technology, evolution, posthumanism, and cooperation.

There are five focus areas of research:

- Space exploration (history, program agenda, habitat architecture, human factors)
- Technology (*theory*: how it mediates our relationships with the environment, how it evolves, how it transforms humankind, how interventions are experienced; *application*: which technologies will be relevant for life in Space, which ones will best serve our stated objectives, and which will most likely be viable)
- Adaptation (the conventional biological processes which are physiological but also incorporating phenomenological processes which are psychological—in other words, how we perceive, experience, and cognitively process our environment)
- Paradigms of evolution (natural selection/Darwinism, symbiotic theory, posthumanism)
- Cooperation (in the context of symbiotic systems, of evolution, and of structures of governance)

Three research areas in particular—Space exploration, emerging technologies, and evolution—all demand a futures perspective. However, there is no single, all-encompassing futures perspective that is relevant for all three areas. Trajectories for advances in Space exploration, which has been sluggish over the last few decades (especially in contrast to its revolutionary beginnings) are slower to develop than those of technology, which has been accelerating even more rapidly in the last few decades. Evolution, by definition, is change over time, which implies a much longer duration than the latter two.

Likewise, there is no one-size-fits-all perspective for different timescales in the future. Human migration to Space, as mapped out in this dissertation, straddles three different timeframes: the immediate future (10–20 years out), the near future (20–50 years out), and the distant future (50+ years). Different issues arise within each of these timeframes: conditions for the first visitors/tourists/pioneers to Space—who are there on a temporary basis—will not be the same ones for 'lifers'—those who will reside there permanently, nor will the first generation emigrants be dealing with the same issues as subsequent generation natives.

The strategies we pursue to both dwell in Space and endure as a species are not just matters of practicality and survival, but also have profound political ramifications. While the ostensible objective of the dissertation is to develop the soundest means to insure long-term adaptation beyond the confines of Earth, it also prescribes a way of interpreting and living in a universe that is a marked departure from the present status quo. Our modern patriarchal values and behaviors—deeply entrenched in a Darwinian worldview that is based on Hobbesian tenets of competition, superiority, and conquest—lead to social practices that are ultimately destructive. However, a shift in this traditional worldview—predicated on recent discoveries in the fields of symbiotic evolutionary theory and cybernetic theory—is already underway. These new revelations in science—in changing our understanding of the world—are beginning to sway our socio-political relations in favor of mutualism and collective alliances through integrative and cooperative practices that encourage mergers and fusions, over the current divisive, unilateral attempts towards individual domination in order to 'get ahead.' Principles derived from symbiotic theories offer an alternative from which to develop new policies that can be applied not only to our impending presence in Space, but also towards our existing destabilized socio-political structures here on Earth. An approach based on reciprocated altruism can serve as a catalyst for positive widespread changes in our relationship to the environment—including both the living and non-living entities that comprise it—as well as inform the methods of technological implementation in the process.

While the range of political issues associated with human migration to Space is too broad to be comprehensively addressed in this project, the dissertation does specifically incorporate some salient political and ethical themes entailed by the prospects of transplanting humans to Space. As humans move into extraterrestrial territories, what protections or rights should be extended to those environments? What kind of social organizational structures will lend themselves to creating the most robust seminal Space communities in order to insure the goal of long-term adaptation? What are the ethical questions regarding the various entities such as future humans in Space, the increasingly intelligent robots that will accompany them, and the possible extraterrestrial life forms we may eventually encounter?

1.6 Methods and Format

Since a major component of the dissertation is based on evaluating different technological approaches for human habitation in Space, methods involving technological assessment are the most appropriate. There are several methods which are suitable, including normative and exploratory forecasting, technology sequence analysis, relevance trees, morphological analysis, mission flow diagrams, and alternative scenarios. Methods with a technology focus are also popular with government high-tech organizations such as the military, DARPA, and the Space program. This dissertation will employ both normative and exploratory forecasting methods, in conjunction with morphological analysis.

Normative forecasting, as indicated by its name, is a prescriptive approach that is applied when specific outcomes are desired. Starting from a future end and progressing inwards, designers and end-users must first define a technological objective to be met and then assess the possible means by which to attain the desired goal. The subjective point of view is fundamental to this top-down method, and therefore incorporates value-laden terms such as 'ought,' 'should,' and 'prefer.' Unlike normative forecasting, exploratory forecasting starts in the present and looks outward, determining what kind of technologies will emerge in the future. This bottom-up method adopts a positivist perspective that focuses on how things are or will be rather than how they should be. There is no element of preference or value associated with it. The normative forecasting method must have an exploratory component however, in that it must be grounded in an understanding of all the available options, as well as the feasibility and the benefits/drawbacks of each, before it can recommend a particular course of action. Since adaptation in Space will require the use of intensive and diverse technologies, I offer four potential implementation scenarios, or approaches, based on our intentions towards the extraterrestrial environment.

As the primary goal of this project is to develop an approach that will best insure long-term adaptation in Space, the focus is on prescribing specific technological *strategies* that lead to long-term adaptation rather than specific evolutionary *outcomes* themselves. What type of beings we become is not a driver of the research. There is no vision or preferred outcome for what humans should become—only how we should arrive there. At this point the dissertation reverts to exploratory forecasting to study the effects our interventions will have on human evolution.

There are several obstacles to achieving the goal of long-term adaptation that I identify and address throughout the dissertation:

First, adaptation is erroneously equated with habitation; a more comprehensive understanding of adaptation must include an evolutional perspective which is overlooked at present. The conditions for long-term adaptation need to be first identified and recognized in the Space industry as a priority for design research and development.

Second, even the current habitability design criteria are insufficient to achieve its more modest goal in that the model lacks a phenomenological perspective. This can be attributed to the fact that the contributions of social scientists are still considered extraneous. Social science needs to play a larger role in habitat design.

Third, our prevailing Darwinian worldview hinders the processes of adaptation in that it favors strategies of competition and conquest that are inherently destructive, over strategies of cooperation and collaboration that are inherently productive.

Fourth, our present tendencies in the way we implement technology with regards to the environment do not bode well for our future in Space. Technologies are customarily used either for military purposes, resource exploitation, or economic gain, but not explicitly with the objective of integrating humans with their environments. This stems from our Darwinian worldview, and needs to change.

The sequence of the dissertation is as follows:

Chapter 2 sets the context of Space exploration, and gives an overview of past, present, and future objectives, including a comparison of the various design criteria they entail. This is followed by a critique of the status quo (the existing model of short-term, habitational adaptation based on physiological, psychological, and social processes) that will identify deficiencies to be addressed and define problem parameters. Solution parameters are proposed through an alternate approach (a new model of long-term, evolutional adaptation based on supplemental phenomenological processes).

Chapter 3 elaborates on the reasons for a phenomenological approach and how this approach will redress deficiencies of the habitational model. It also discusses how technological interfaces that mediate our relationship to the environment are central to this approach and critical to achieving long-term adaptation.

Chapter 4 begins with an overview of theories of technology as grounding for the subsequent discussion on specific technological interventions for human dwelling in Space. It then introduces a range of alternative technological approaches (avoidance, domination, subordination, integration) that are informed by our attitudes towards the environment.

Chapter 5 presents extensive justification for the integrative approach through a theoretical grounding in evolutionary theory, environmental theory, and theories of cooperation. Cooperative strategies for specific technology applications as well as for social organization in Space are proposed.

Chapter 6 examines the implications of both migration and technological intervention for human evolution in Space, beginning with how processes of adaptation transition to processes of evolution. Modes of human evolution are discussed through various timelines in the future, each of which presents a different set of issues. The chapter concludes with reflections on what it means to be posthuman.

References

Gardner, J. N. (2003). *Biocosm: The new scientific theory of evolution*. Maui, HI: Inner Ocean Publishing.

Hayles, N. K. (1999). *How we became posthuman: Virtual bodies in cybernetics, literature, and informatics* (1st ed.). Chicago, IL: University of Chicago Press.

Chapter 2
Current Directions in Space Exploration

2.1 Premise

The purpose of this chapter is to set the context for the dissertation by giving an historical overview of the past and present goals of the exploration of Space and the design criteria associated with each. With the more recent objectives of establishing a human presence on Mars as the next major milestone in exploration (followed by human settlements), a new set of design challenges has emerged. In response to those challenges, human factors research has received increasing attention, considerably expanding in scope over the last two decades to now address requirements for habitability. However, despite the latest focus on habitability criteria, most current habitat design proposals are still rooted in a rigid engineering perspective and not in a deeper understanding of what promotes well-being; to remedy those oversights, habitability criteria must be re-evaluated.

But even this broader scope of habitability is still insufficient, in that habitability is concerned only with matters of well-being and comfort; and while certainly a critical aspect, this definition is an oversimplification of what it means to habitate, as well as to inhabit a place in particular. As such, the current meaning ignores longer-term issues like the development of resilience, robustness, and durability. These are glaring deficiencies in the current approach which also require more serious attention.

2.2 The Changing Goals of Space Exploration: Past, Present, and Futures

The launch of Sputnik in July 1957, officially ushered in the Space era, demonstrating the Soviet's technological superiority at the height of the Cold War. The U.S. feared that this breakthrough would eventually allow the Russians to expand their domain of sovereignty into Space, posing a threat to the U.S.'s sense of

E.S. Lockard, *Human Migration to Space*, Springer Theses,
DOI 10.1007/978-3-319-05930-3_2, © Springer International Publishing Switzerland 2014

identity and their place in the global arena. President Kennedy reacted by putting an ambitious challenge to our country: send a man to the Moon and deliver him back safely within the decade. Though we were nowhere close to having this technology, the public rallied behind what was touted as a noble scientific and technological pursuit, but the motivation was purely political. Despite skepticism on the part of the enjoined engineers, the endeavor resulted in one of the most heroic achievements in the history of humankind.

After several previous attempts, Apollo 11 successfully landed on the Moon in July 1969; six more Apollo missions would follow, after which national interest in returning to the Moon began to subside. As the industry turned to the development of large orbital Space stations, Space exploration became more of a collaborative effort between nations, particularly between the U.S. and Russia. When flying into Space was no longer an engineering 'feat,' the focus shifted to scientific research of the solar system through the launching of several probes. In 2004, in order to revitalize the waning Space program, President Bush announced the Vision for Space Exploration. NASA's Constellation program was established, and set its sights on establishing human outposts on the Moon and later on Mars between the 2020s to late 2030s. Since then, several rovers have been dispatched to Mars with the specific objective of studying conditions on the Red Planet and searching for signs of past or present life. The presence of methane (a volatile gas that is produced either by volcanism or by microbial life) as well as water-ice at the poles has raised hopes for the discovery of past, or maybe even present life there.

NASA and the Russian Federal Space Agency (RKA; also known as Roscosmos) no longer have the monopoly on Space exploration they once enjoyed decades ago. Other Space initiatives are gaining momentum; in particular, China National Space Administration (CNSA), Japanese Aerospace Exploration Agency (JAXA), the European Space Agency (ESA), and India Space Research Organization (ISRO). In addition to these programs, over a dozen other countries have already developed extensive launch capabilities, including Canada Space Agency (CSA), Iranian Space Agency (ISA), Korea Aerospace Research Institute (KARI), Israeli Space Agency (ISA), Italian Space Agency (ASI), French National Center of Space Research (CNES), the United Kingdom Space Agency (UKSA), National Space Agency of Ukraine (NSAU), and Indonesia National Institute of Aeronautics and Space (LAPAN). Only four however, have manned spaceflight capability at this time: RKA, NASA, ESA, and CNSA. Getting to Mars holds special fascination for many of these programs: besides the U.S. and Russia—who have already traveled to Mars—China, France, and Japan have also attempted (albeit unsuccessfully) to reach the Martian orbit; India is currently in preparations to make an attempt later this year.

Despite the surge in national programs, it is very possible that the greatest strides in the near future of Space exploration may instead come from the private sector, where an abundance of funding from wealthy benefactors, coupled with capitalist ambition, are accelerating those target objectives. Virgin Galactic, brainchild of visionary entrepreneur Richard Branson, offers trips to LEO, and is planning to expand after initial success and additional private investment.

The Mars One Project is accepting applications for a one-way trip to Mars in 2023, to establish the first permanent human settlement there. All aspects of the venture will be broadcast around the world in reality TV format which will serve as the basis for raising project funds.

Inspiration Mars, founded by billionaire Dennis Tito, and in conjunction with Paragon Space Development Corporation (founded by Jane Poynter and Taber McCallum of the Biosphere 2 experiment) are likewise seeking candidates, specifically married couples, to send on a 500-day flyby mission to Mars and back in 2018. SpaceX (Space Exploration Technologies) was founded by entrepreneur Elon Musk, and has been contracted by NASA to develop several launch vehicles for its upcoming missions.

2.2.1 Missions (Short-Term Duration) vs. Habitation (Long-Term Duration)

In the earliest missions to Space, the singular human factors criterion for the design of manned spacecraft was the safety and survival of the crew—getting a human into Space and back to Earth alive. Later during the 1970s–1990s, as missions to Mir, Skylab, and the ISS took on a scientific agenda, conditions related to work performance received increased attention. During this period of short-term missions beyond Earth, the realm of human factors pertained only to the physiological effects of prolonged stays in Space. However, as we embark upon a new phase of Space exploration—the goal of which is to establish permanent habitable outposts on the Moon, Mars, and nearby asteroids—the scope of human factors must expand to take into consideration psychological and social aspects, especially since they will become increasingly more crucial to the crews' survival in long-duration voyages. The importance of non-physical factors was stressed by Russian scientist Dr. Oleg Gazenko, when he stated that "the limitations of life in Space are not medical, but psychological" (Stuster, 1996, p. 165).

Throughout history, voyages of long duration and extreme hardship—both on Earth as well as in Space—have demonstrated our species' capacity for adapting to dangerous conditions and harsh environments. But where the primary objective of exploration has been the acquisition of scientific information, the objectives of habitation are fundamentally different. Psychosocial concerns will take on a much higher priority than they did during short-term voyages, where it was feasible for the crew to suppress their less urgent needs for the duration of the mission. The need for a new human factors focus became apparent in a study involving the design of a lunar laboratory where "it was concluded that the primary limiting factor for all extraterrestrial activities is the 'problem of interpersonal relationships' among the isolated crew (La Patra, 1968)" (Stuster, 1996, p. 165).

As long as the needs for survival are met, almost any other condition can be tolerated for a short period of time. However, as the time duration is extended, certain needs can no longer be ignored. If not adequately addressed they can even undermine the success of the mission. But exactly at what point in time do other needs become essential? There is no clear-cut threshold between what is defined as short-term and what is defined as long-term. Short-term missions last from a few days up to a few months, but are typically measured in weeks. Long-term durations refer to periods of time greater than 6 months, and are usually framed in terms of years. Mid-length durations span anywhere in between, and is therefore the most challenging to define. This dissertation is concerned only with very long durations, consisting of years at minimum, and even generations—a timeframe that falls outside the scope of current design objectives. Given the enormous implications of human migrations to Space—both to humans and the Space environment—a longer-term perspective should be part of the vision.

2.2.2 Paradigm Shift in Human Factors Research

With the recent initiative in the Space industry to eventually establish colonies on Mars and possibly in low-Earth orbit (LEO), there has been increasing recognition that the requirements for trips of long duration—as opposed to those of short-term voyages—need to take into consideration issues of comfort (Wapniak et al., 2007). This broadening of human factors parameters entails a fundamental shift from the quantitative challenges of insuring physical survival (e.g., how much oxygen is required to survive?) to the qualitative challenges of insuring overall human well-being (e.g., how can the deleterious psychological effects of isolation be mitigated?) Rather than focusing exclusively on how humans *endure*—the criteria for which are based on designing for tolerances—the emphasis has shifted to how humans *thrive*—the criteria for which are based on designing for optimization. "Narrow views of human factors limited to anatomy, physiology, biomechanics, perception and cognition are giving way to enlarged views that include emotions, attitudes, personality, interpersonal relations, group and organizational dynamics, and culture. Contemporary human factors extend beyond work to self-maintenance (sleeping, eating, personal hygiene) and leisure time activities" (Harrison, 2010, p. 891). Recognizing the need for more than just the bare necessities for survival, the former paradigm—known as the "Spam in a Can" model, which assessed *minimum* requirements—has been gradually replaced by a "Quality of Life" model, which assesses *adequate* requirements (Dudley-Rowley et al., 2003).

The notion of habitability, essential to the "Quality of Life" model, includes criteria for safety and security as its basis, but goes well beyond by focusing on human well-being and comfort (Fig. 2.1). NASA has defined habitability as "a measure of the degree to which the environment promotes productivity and well-being" (Preiser, 1991, p. 150). Within this definition there are three levels. The first level is health, safety and security. This level takes the highest priority and must precede all other considerations. In its nascence, the Space program addressed only this criterion in missions of short duration. The second level is functional and task

TIMELINE:	Short-term missions (1970's—1980's)	Long-duration voyages (1990's—2010's)	First Mars migration (2020's—2030's)
DESIGN CRITERIA:	survival	performance	habitation
SCOPE OF TECHNOLOGIES:	safety (physical)	capabilities (functional)	well-being (comfort)

Fig. 2.1 Progression of design objectives in human Space exploration

performance. This level rose in significance when scientific research and experimentation became a more integral component of the mission. Mission durations also increased from days in orbit to weeks or months on the space stations. The third level of habitability is psychological comfort and satisfaction.

The third level for habitability was not considered crucial in short-term missions, but has grown more important in missions of longer duration, as psychological issues can undermine both work performance and crew morale. And when psychological or social problems are not adequately addressed they can even threaten the very survival of the crew, should social cohesion disintegrate to the point where members of the team can no longer cooperate with or depend on each other. Preiser advises: "Of greatest concern to programmers and designers should be the third level of habitability, namely, the short-term and long-term impact of space settlements on the psychological comfort and satisfaction of their occupants. This aspect of environmental design has been largely ignored in recent history, as the sad condition and image of our built environments demonstrates. The repetition of inhumane, discontinuous, and hard environmental design that we commonly face today should be avoided at all costs in space stations and settlements" (Preiser, 1991, p. 154). However, despite this acknowledgement, living quarters are still designed only for survivability, with any amenities treated as frivolous embellishments rather than as necessities. "All too often there is a serious neglect of both architectural and behavioral issues in the course of mission planning, an oversight that encourages the development of little more than inhabited machines. This is a false economy because overlooking the human element increases rather than decreases costs" (Vogler and Jorgensen, 2004; Harrison, 2010, p. 895). This criterion must become a paramount consideration in establishing human settlements in Space.

While human factors concerning health and safety can be quantified, and productivity can be measured in output, well-being is a more nebulous concept to pin down. From NASA's engineering-dominant perspective, it can only be understood in observable, functional, or behavioral terms—terms that do not acknowledge interpretive, subjective experience which is also essential to individual well-being. What is well understood by architects, interior designers, environmental psychologists and other social scientists is that how humans *perceive* their surroundings and ascribe meaning to them is a critical aspect of habitability—as much if not more so than the actual conditions of the environment. Space psychologist Peter Suedfeld confirms this:

> Researchers should think in terms of experiences within environments rather than of environmental characteristics. The relations between environmental features and behavior must be studied in terms of interaction, not mainly effects. As several investigators have pointed out, the environment has no direct impact on human beings. Rather, it is filtered through their psychological and physiological information-processing systems. In consequence, the crucial determinant of the response is not an environment, but an experience, this being defined as the environment *and* its meaning to the individual. Researchers should, therefore, adopt some new ways to gain an understanding of environmental impact on people. The most obvious one is to measure not only how individuals behave in the environment but also how they perceive it (Suedfeld, 1991, pp. 137–138).

Jack Stuster supports Suedfeld's distinction between experience of the environment (subjective) and the physical features of the environment itself (objective). Referring to his personal experiences from his polar expeditions, he cites the importance of views: "[There are] strong preferences for nature scenes, particularly seascapes and scenes involving wide-open spaces. . .a desire to view landscape-type scenes and simulate the exercise of distant vision." In response to this yearning, "false windows have been installed in personal quarters and common areas to display various types of dioramas. . .reportedly, some individuals enjoy staring at the dioramas for long periods of time" (Stuster, 1996, p. 191). The experience of Nature does not have to rely on actually re-creating it, but can be elicited through representation.

So what exactly are the conditions that promote "psychological comfort and satisfaction"? This is the area of human factors research that is the most difficult to quantify, as it has more to do with human perception of the environment (subjectivity) than it does with the actual environment itself (objectivity). Habitability criteria have been evaluated by several Space architecture programs in academia, where most of the criteria focus on the experience of the occupant, rather than the environment itself. They can be categorized as follows:

Perceptual/Experiential:

- personal identification
- sense of belonging
- familiarity
- unpredictable conditions
- sensory stimulation
- sensually pleasing
- egalitarian conditions
- privacy
- variety
- comfort
- playfulness
- leisure
- contemplation
- flexibility
- stability and security
- social interaction

Environmental:

- contact with nature
- regulation of light
- views
- orientation
- functionality
- acoustics/environmental sound

Other research in environmental psychology concluded that four essential qualities are the basis of preferred living environments, in which subjects feel competent, confident, and in control: legibility (clarity of individual spaces and the circulation between them), coherence (total relationship between composite spaces), complexity (variety and diversity of spaces, density of information), and mystery (sense of surprise, discovery) (Kaplan, 1988). All of these qualities relate to subjective human perception, and prescribe nothing specific about the architectural features of the living environment itself.

2.3 Environmental Challenges of Space

The challenge of sending unmanned vessels to Space was one of physics; the challenge in sending manned flights is one of physiology; and the challenge of sending humans to live in Space is psychological. In order to develop more appropriate design strategies for habitation in Space, the range of stressors, from the physiological to the psychosocial, must be identified and systematically addressed.

2.3.1 Physiological Stressors

The main issues in the category of physiological environmental stressors include:

2.3.1.1 Prolonged Exposure to Micro- or Partial Gravity

Short-term effects (lasting up to one week) include spatial disorientation, facial edema and redistribution of bodily fluids, nausea, increased heart volume, increased urination, dehydration, and digestive disturbances. Also known as Space Adaptation Syndrome, most of these effects are temporary, and subside or disappear after only a few days (Cagle, 2009). Long-term effects however, are more severe and degenerative. Those include muscle atrophy, anemia, decalcification (which can also lead to kidney stones), bone density loss leading to osteoporosis, and heart atrophy. These ailments do not taper off but increase with time, making longer durations in Space a greater physiological risk (Kanas and Manzey, 2008).

Both short-term and limited long-term effects associated with micro-gravity have been studied on the Space Shuttle, the International Space Station, and on the "Vomit Comet" (which simulates a zero gravity environment through parabolic flight), but because precise simulation on Earth is not possible—underwater simulations are the closest we can provide—the effects of longer exposures cannot be thoroughly studied until we are actually living in Space.[1]

It is generally believed that in order for humans to survive the long haul in Space, the provision of some kind of gravitational force will be necessary. Ted Hall and Olga Bannova have authored a number of papers that propose artificial gravity environments in LEO venues. Rotating habitats around a fulcrum generating centrifugal force can simulate terrestrial gravity conditions; the strength of the gravitational force depends on the velocity of rotation and the length of the rotating arm. The most serious drawback to this concept is the coreolis effect, in which an additional force due to the momentum of the swinging armature affects motion within the interior. This can be reduced by modifying the variables that also determine the amount of gravity produced. It may also be possible that humans can acclimate to the presence of this dynamic effect, but this has not yet been studied.

2.3.1.2 Ionizing Radiation

Shielding from ionizing radiation (cosmic radiation and solar flares, as well as trapped radiation belts) will become a necessity in Space, where there is neither a protective atmosphere to filter out harmful rays, nor a magnetosphere to repel them. Moreover, because the types of radiation are different—solar flares are intermittent but intense bursts of radiation from the Sun, while cosmic radiation is omnipresent and diffuse—design requirements for shielding will vary. Too much exposure to ionizing radiation—the effects of which are cumulative—destroys the body at the cellular level and leads to increased incidence of cancer and immunodeficiency-related illnesses, thus making radiation the most life-threatening of the physical stressors.

The most efficient and economical solutions that have been proposed so far would be to either cover a lunar habitat with several feet of regolith (lunar 'soil'), or set up dwelling in subterranean caves and lava tubes. However, while these solutions may be the most expedient in terms of both cost and safety, there are drawbacks when evaluating the proposal from an experiential (as opposed to a

[1] A macro-g environment—which could in fact be partially simulated in a terrestrial context—will present even more challenging conditions than micro-g, but since this scenario is unlikely for the near future, the scope of the dissertation will be restricted to the problems associated with micro and partial gravity only.

purely practical) perspective. By reinforcing the barrier between inside and outside, occupants will be cut off not only physically, but also visually and acoustically from their surroundings, and are hence more likely to feel an exacerbated sense of incarceration as well as a disconnection from their environs as a result.

2.3.1.3 Diurnal and Orbital Cycles

Diurnal cycles vary greatly depending on the extraterrestrial venue. On the Moon, one day/night cycle is equal to approximately 28 Earth days (14 days of light and 14 days of darkness). Compared with Mars, the 'sol' (Mars' diurnal unit of time) is—at 24 hours and 37 minutes—slightly longer than Earth's day. This difference is negligible and thus entrainment to the Mars cycle would be almost effortless; however, communications with Earth would become gradually desynchronized. In contrast to planetary venues, the light of the Sun in low-Earth orbit would be ever present, since there would be no planetary rotation to establish a diurnal cycle. In Arctic and Antarctic simulations where sunlight was present for up to 23 hours a day in the summer months, the quick and easy solution was to block the external light with window shades at bedtime in order to simulate nighttime conditions conducive for sleep.

Since humans' circadian rhythms evolved according to Earth's 24-hour diurnal cycle, life in Space—with the exception of Mars—will require some kind of environmental cues, or zeitgebers, incorporated into the habitat to insure that the crew's circadian rhythms, which determine sleep and wake cycles as well as mood, are kept synchronized. Insufficient or poor quality of sleep is well known to be detrimental in numerous ways: both performance and productivity are compromised by fatigue, forgetfulness, and inaccuracy. Deficiency can cause irritability and thus adversely affect social relations. Lack of sleep and excessive exposure to light also inhibit the body's production of melatonin, the chemical which regulates mood. There are also physiological ailments associated with deficient melatonin production, such as a higher incidence of certain types of cancer (Kloog et al., 2010; Kantermann and Roenneberg, 2009). When survival depends on both maintaining social cohesion as well as work performance, insuring adequate sleep is essential. The design of the habitat should employ internal and external measures to regulate appropriate cycles of sleep and wakefulness.

2.3.1.4 Magnetic Fields

Little is known about how geomagnetism impacts human behavior and well-being. But all life on Earth has evolved in the presence of its geomagnetic fields, and like Earth's diurnal cycles, these forces "help regulate the biological cycles of many species" (Gallagher, 2001, p. 84). There is evidence that the "behavioral cycles [of organisms] are normally timed by cyclical changes in the strength of the geomagnetic field as the Earth rotates. Like light, the Earth's field may keep living things in sync with the solar system" (Gallagher, 2001, p. 84). While options to generate an artificial magnetosphere have been proposed as a possibility for shielding spacecraft against cosmic radiation and solar flares, testing indicates they are only effective for repelling SPE's and are inadequate for protection from GCR (Seedhouse, 2009).

The scant research on magnetic fields has been limited to studying shielding capabilities, and little has been mentioned about the potential benefits for human adaptation in Space. This is an area that calls for further study, but given the status of the technologies at present, it seems unlikely to become viable in a foreseeable future.

2.3.1.5 Other Factors

Other physical stressors include the need for pressurization (both in LEO and on planetary sites where there is little or no atmosphere);[2] air composition (absence of toxic gases, optimal amounts of oxygen intake, and discharge of expelled carbon dioxide); and sub-freezing temperatures.[3] These environmental stressors have serious physiological effects on all the systems of the body: the cardiovascular system, the vestibular and sensory/motor system, the musculature-skeletal system, sleep and circadian rhythms, and cognitive processes and performance (Kanas and Manzey, 2008). Extensive research has been done to study these effects and search for ways to resolve them. Since most of the conditions beyond Earth exceed the body's somatic limits for adaptation, technological mediation will be necessary in greater or lesser degrees.

2.3.2 Psychosocial Stressors

These include a number of non-environmental factors, such as crew composition, crew size and number, degree of heterogeneity or homogeneity, and duration of trip. Although much social research has been done in the area of crew dynamics, the focus of the dissertation is on environmental factors.

In terms of long-term habitability, the most challenging psychosocial stressor in Space arises from the isolation and confinement imposed by the remote environment. While one condition generally accompanies the other, there are distinct differences between the two states that require specific interventions. Isolation is a psychological state caused by physical separation or phenomenological distance from one's social network or sphere of familiarity. Confinement is a physical state

[2] Without sufficient external forces of pressure, the body's internal fluids would vaporize.

[3] Due to either a lack of atmosphere to trap warming greenhouse gases or distance away from the Sun.

in which either personal space is limited or one's mobility is restricted (Evans et al., 1988).

Typical problems associated with long durations spent in isolated and confined environments (ICEs) include sensory deprivation or overload, loneliness, anxiety, depression, and boredom. Other psychological effects include feelings of helplessness and loss of control, perceptual narrowing (solipsism), and claustrophobia; in the most extreme cases, psychotic episodes can even occur. These in turn have consequences for social interaction: development of aggression and hostility towards others, amplification of interpersonal differences, territorial behavior, withdrawal from the group, stimulus-seeking or provocative behavior, and group splintering and polarization (Kanas and Manzey, 2008). Distorted group dynamics and irrational behaviors were cited by crew member Jane Poynter as one of the major problems that occurred during the Biosphere 2 experiment (though the recent Mars 500 simulation did not seem to suffer from the same difficulties). There are concomitant physiological effects as well, such as increased ailments, sleep disturbances, and lethargy. Not surprisingly, because of these negative side effects, the resulting decrease in motivation and alertness can impair work performance. "In the well-known 'kitten experiment,' the inability to move prohibited one kitten from becoming perceptually alert and from functioning normally. By analogy, it is feared that the envisioned space stations and settlements may limit human movement to such a degree that similar perceptual impediments may occur" (Preiser, 1991, p. 159).

In addition to imposed isolation and confinement, extreme environments entail a lifestyle marked by polarized conditions. On one end of the spectrum such environments are characterized by remoteness, solitude, small crews, monotony, tedium, boredom, and a paucity of other types of life forms—all the makings of an under-stimulating environment. Daily tasks and responsibilities—though they may be essential to the functioning of the outpost—are typically mundane and repetitive, with little variation, spontaneity, or unpredictability. On the other end of the spectrum, even small accidents or careless error could threaten the very survival of the entire crew; thus there is a perpetual sense of peril and risk that constantly reminds the crew how tenuous their circumstances are. These highly stressful situations create an environment of over-stimulation, which can then cause agitation and anxiety.

In small doses, both under- and over-stimulation can have beneficial effects on crew members. Brief periods of low stimulation can be soothing and relaxing, allowing time for greater reflection and introspection. Memory is improved and there is often an increase in creativity as well as positive behavioral changes. Likewise, high stimulation in short durations of time is invigorating and exciting, while also increasing alertness. However, in longer durations, both under- and over-stimulation can become psychologically toxic. Prolonged exposure to under-stimulating conditions leads to lethargy, boredom, and depression. It can also cause more extreme reactions such as hypochondria, psychosomatic illness, and excessive introspection. "Inhabitants of a contained environment lack the variety of sensory input and diversity of motor activity that are important for maintaining optimal thinking, feeling, and behavior in the human" (Blair, 1991, p. 58). "An overly long stay in an under-stimulating environment poses problems...in an under-stimulating milieu, we subconsciously try to stir up some action from within" (Gallagher, 2001, p. 146). Instead of inducing positive behavioral changes, prolonged under-stimulation can result in behavioral breakdowns which disintegrate social cohesion. People suffering from under-stimulation are also highly suggestible and grow increasingly solipsistic, no longer capable of acknowledging an external reality beyond the confines of their immediately perceivable surroundings.

The effects of extended exposure to over-stimulating conditions are similarly deleterious. Excessive over-stimulation tends to be accompanied by feelings of anxiety as well as a tendency to be less helpful to others. Moreover, "over-stimulation provoked by crowding can lower mood, increase withdrawal and aggression, and decrease helpfulness, particularly in men" (Gallagher, 2001, p. 185). Therefore, it is critical for both personal well-being as well as social cohesion, to find a balance of high and low stimuli in an isolated and confined environment.

Although there has been an abundance of research examining the effects of isolation and confinement, there has not been sufficient investigation of how the physical design of the habitat itself contributes to or can ameliorate those problems. Proposed solutions thus far have included: improving communications with distant loved ones, testing potential crew members for individual tolerance to stressful conditions (rather than training them to better handle stresses, like the Russians do), and intra-crew dynamics (which tends towards homogeneity in crew composition in terms of gender, profession, and age). With the notable exception of provisions for privacy and personal space, virtually nothing else has been proposed in the way of architectural strategies.

Solutions for the challenges of physical survival, though perhaps difficult to implement, are usually more straightforward than the challenges posed by behavioral problems, whose factors may be more complex and layered. If there is not enough oxygen for the body to function properly, for example, the effects are immediately evident and easily attributable. The solution is simply to provide more oxygen. But the psychological and social issues are not as easy to identify or remedy in the same quantifiable manner. For example, how much space is necessary to avoid feeling crowded? This is not strictly an issue of physical conditions, as factors such as gender, cultural norms, and group dynamics play a significant role in feelings of crowdedness (Gallagher, 2001). How should the limited amount of space be configured to afford adequate privacy? Privacy (being away from others) and personal space (the sense of territory and domain of control) are likewise influenced by social factors, and not just physical boundaries. How can the design of the habitat diminish cognitive disengagement from the outdoors, which can then lead to solipsism? Ultimately, these are questions of optimization, relative to a range of specific constraints and not an absolute condition. Social science research, such as analog environments and simulation studies are the only way we can test possible solutions to many of these questions.

2.4 Research on Space Habitation

Human factors research in Space exploration covers a broad spectrum of disparate topics. The architectural design of the habitat itself is often narrowly interpreted as its engineering requirements such as payload and thermal shielding, but it also includes other practical human-related considerations, such as anthropometrics, ergonomics, space planning, functionality, work performance, and comfort. This area is the most thoroughly studied, with contributions from Scott Howe, Marc Cohen, Kriss Kennedy, Brent Sherwood, Constance Adams, Rachel Armstrong, Barbara Imhof, Susmita Mohanty, Martina Pinni, and students from a number of academic Space Architecture programs. Ted Hall and Olga Bannova have written several papers on artificial gravity systems, both from an engineering perspective as well as how it affects human occupants.

Psychological and sociological aspects of living in Space are a more recent field of study. A new field of astrosociology has also emerged that introduces the perspective of how Space exploration impacts terrestrial society. Research in the psychosocial area covers a wide range of topics, such as proxemics (spatial spheres of social interaction), crew composition, crew dynamics, salutogenesis (health-promoting effects of high stress situations), behavioral pathologies, the influence of culture, etc. Major contributors in these areas of study include Marilyn Dudley-Flores (formerly Dudley-Rowley), Thomas Gangale, Jim Pass, Sheryl Bishop, Kim Binsted, Nick Kanas, Dietrich Manzey, Peter Suedfeld, Jack Stuster, Eva Ihle, Jennifer Ritsher, David Swift, and James Dator.

An area that has received scant attention is the phenomenology of Space, which delves into the subject of how humans perceive and experience their environments, and how those perceptions and experiences impact our relationship to our environment. Probably the most noted research in this area is on the Overview Effect, first documented by Frank White. Other contributors in this area include architectural theorists Andreas Vogler and Jesper Jorgensen, as well as futurists James Dator and Magoroh Maruyama.

2.4.1 Historical Precedents

Historical precedents are limited to a few examples: the ISS, Mir, Skylab, and the Apollo missions. All of the scientific knowledge acquired on how the body reacts to the absence of gravity (in both short and long exposures) comes from studies done on the Space stations. The effects are extensive—affecting the vestibular, sensory/motor, and nervous systems—and long durations in microgravity conditions pose one of the greatest physiological challenges for humans. It is possible that future modules for the Space stations will have a rotating armature to generate artificial gravity conditions that can be tested in addition to other physiological factors.

Anecdotal evidence and testimonies from astronauts have been invaluable in researching psychological and social issues, both between members of the crew (human-human interface) as well as between the crew and the physical enclosure itself (human-technology interface). Nick Kanas and Dietrich Manzey introduced the relatively nascent fields of Space psychology/psychiatry, and mission research from Peter Suedfeld led to his recognition of the salutogenesis phenomenon.

2.4.2 Terrestrial Analogs

There are a number of terrestrial situations that are comparable to aspects of living in Space. Some of the analogs have been incidental, such as submarines and prisons, and others such as Antarctic research stations and Biosphere 2 were developed in part for the express purpose of studying different facets of Space life. Earth-based analogs are not adequate for understanding the human body's physiological reactions to the physical conditions of Space, but they do provide insights into many of the social issues that are associated with remote environments, such as the negative and positive psychological effects of crowding, lack of privacy and personal space, confinement, isolation, understimulation and overstimulation. Social interactions are another subject of investigation, in particular crew composition (heterogeneity, gender, size, and dynamics). Sheryl Bishop, Jim Pass, Thomas Gangale, Marilyn Dudley-Flores (formerly Dudley-Rowley) have focused on sociological issues from a scholarly perspective.

Research comes however, not only from social scientists but also from first-hand accounts: Albert Harrison, Yvonne Clearwater, Chris McKay, and Jack Stuster have written about their experiences on polar expeditions, and Jane Poynter wrote about her participation as a crew member of Biosphere 2.

Another primary objective of Biosphere 2 was to create a completely closed ecological life support system (CELSS) to determine whether a human colony could live entirely from the resources contained within its enclosure (with the exception of sunlight as the only external source of energy). The results of the 2-year study were not promising, but many lessons were learned from some of its ecological failures.[4]

[4] Oxygen fell to dangerously low levels due to the carbon dioxide emitting from the concrete structure; some of the biodomes suffered from pest infestations; and despite having brought in over 3,800 species, maintaining homeostasis was constantly precarious.

Fig. 2.2 Habitat simulations: HDU, FMARS, MDRS, HI-SEAS, Mars500, Mir training module

2.4.3 *Habitat Simulations*

Given that precedents for habitation in Space are few, and the cost of building actual prototypes for testing in LEO is prohibitive, much of the current human factors research relies on numerous simulations, some of which include: Mars 500 in Moscow, MDRS in Utah, FMARS on Devon Island, HDU in Arizona, HI-SEAS on the Big Island (Hawaii), and Mir training mock-up modules (Fig. 2.2). The simulations are employed to study a range of habitation conditions from entrainment to menu fatigue, but the predominant focus has been on psycho-social issues, which include problems associated with crowding and lack of privacy, understimulation and overstimulation, isolation and confinement. The only simulations that test for reduced gravity conditions are the "Vomit Comet", an aircraft that allows the passengers a short period of non-gravity during its parabolic flight, and underwater training pools.

2.5 Review of Space Habitat Design Proposals

Habitat design proposals for Space range from the feasible to the fanciful, and everywhere in between. Habitat typologies can be classified into three categories: planetary/surface habitats, orbital space stations, and macro-scale space colonies. Of the three, the only ones that have actually been built, flown, and tested in-situ are the Space stations and their modular appendages. But Space stations were not designed with long-term habitation in mind, so with the exception of work performance and immediate physical and psychological needs, human factors considerations were kept at a minimum. Most of the current proposals focus on planetary venues, given that the next directive in Space exploration is to return to the Moon and eventually travel to, and settle on Mars. Despite the prototypes which have been developed thus far, to date only one design in this category has actually been landed and tested: the Apollo Lunar module (Kennedy, 2002). But because it was a vehicle intended primarily for transport of a skeleton crew for a brief period of time and not for the purposes of long-term habitation, it is hardly a precedent upon which to build.

2.5.1 *Existing Habitat Designs*

There are several habitat prototypes that have already been developed, meticulously crafted using the latest technologies, but so thoroughly entrenched in pragmatic considerations that lack any vision of how Space dwellers should live. The first generation of space stations in the 1970s and 1980s—Skylab, Spacelab,[5] and Salyut 7—were intended to serve the purpose of advancing scientific research rather than human dwelling in Space. The second generation of stations—Mir in the 1990s and the ISS at the turn of the century (Figs. 2.3, 2.4, and 2.5)—also focused on scientific research, but with crew members staying onboard for longer durations, coupled with the longer active lifespan of these stations, designers could no longer ignore the broader needs of its crew members.

2.5.2 *Prototypes in Development*

Greater attention was paid to the crew's welfare in subsequent projects such as the highly-publicized Transhab developed by NASA architects Kriss Kennedy, Constance Adams, and Marc Cohen (Fig. 2.6). Comprised of a pre-fabricated solid structural core enclosed by a multi-layered inflatable membrane which would significantly reduce payload, the transit module was intended to provide additional living space and amenities for the astronauts aboard the ISS. The interior spaces included three occupiable levels in a radial spatial configuration around the central core. Unfortunately, the project was halted by the U.S. government before it ever had a chance to be deployed.

Some examples of planetary-based habitat designs that have been proposed so far include: a lunar base membrane structure, a thermal protection concept for a lunar base, an inflatable lunar shelter, and a foam-rigidized inflatable shelter (Imhof, 2003). The Lunox First Lunar Outpost, developed by NASA in 1993, was designed to be a smaller vehicle in order to be more cost-effective by utilizing lunar regolith (ISRU) to produce fuel for the return trip to Earth (Fig. 2.7). NASA is

[5] As part of the Space Shuttle, it was properly a transit hab and not a station per se.

Fig. 2.3 Skylab

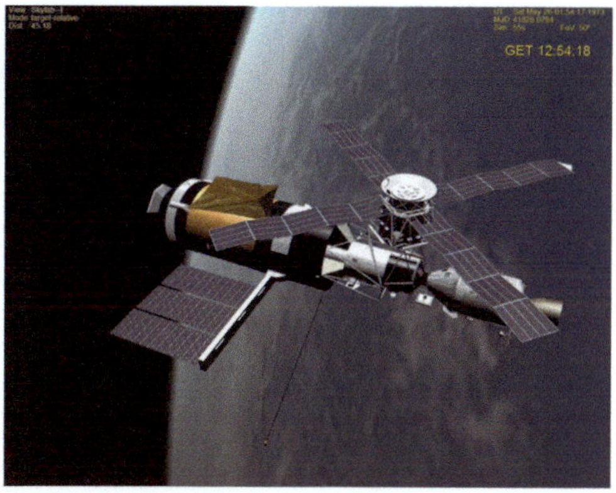

Fig. 2.4 Mir Space Station

Fig. 2.5 ISS

currently developing the HDU (Habitat Demonstration Unit) which is being field-tested this year in Arizona (Fig. 2.8). It is comprised of a solid structure base with an inflatable compartment above for crew quarters, and separate appendages for air lock/dust mitigation, as well as for hygiene.

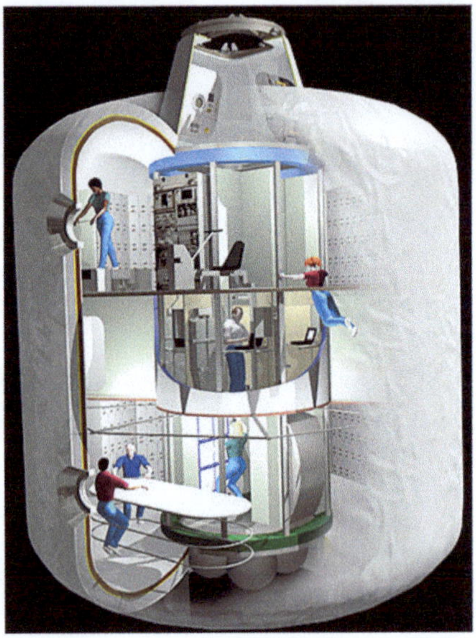

Fig. 2.6 Transhab

Fig. 2.7 Lunox

Fig. 2.8 HDU

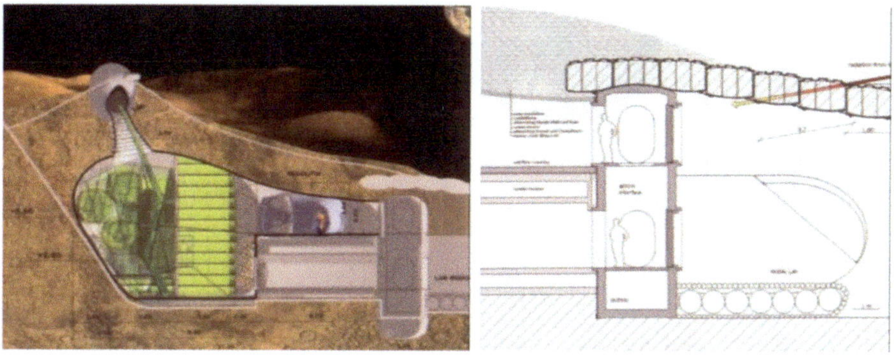

Fig. 2.9 Kepler lunar habitat

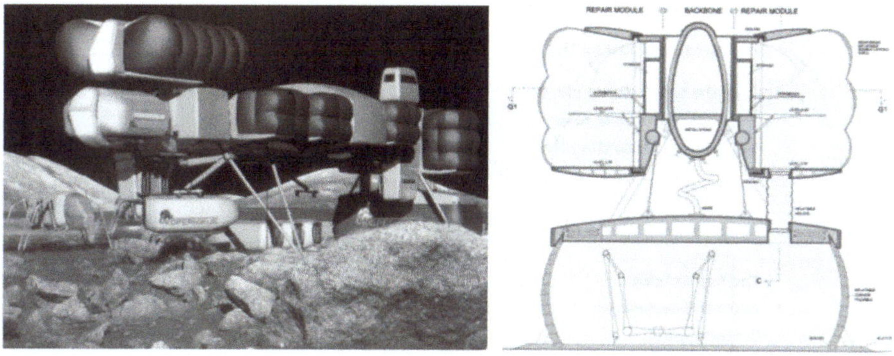

Fig. 2.10 Kopernikus lunar habitat

In addition to the projects being developed at NASA, there are a number of design proposals that come from outside government space agencies, such as the Kopernikus lunar habitat and the subterranean Kepler lunar habitat (Figs. 2.9 and 2.10). Note that only Transhab illustrates the design from the user's interior perspective.

Most proposed planetary habitat designs are either circular (planometrically) or spherical/cylindrical (volumetrically) in shape because this form is the easiest to construct, the easiest to pressurize, the most efficient in terms of structure (therefore using less materials), and provides the maximum amount of internal volume relative to the surface area of the envelope. Interior layouts that attempt to incorporate orthogonal elements have proven to be awkward and inevitably result in unusable (and thus wasted) residual spaces. Radial space planning about a central core is the most logical organization for a cylindrical or spherical typology, since its form is predisposed to rigid symmetry based on partitioning of structural members (Fig. 2.11). But other more optimal internal configurations are possible. Rather than have all areas be equal in space and uniform in shape, configuration of various areas

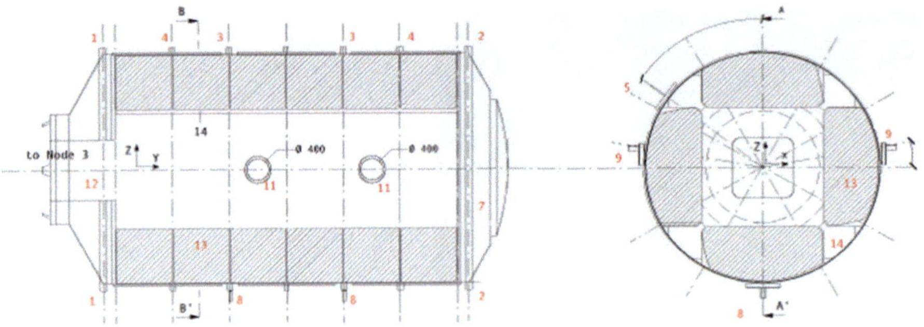

Fig. 2.11 Typical symmetrical 4-standoff module with orthogonal interior volume

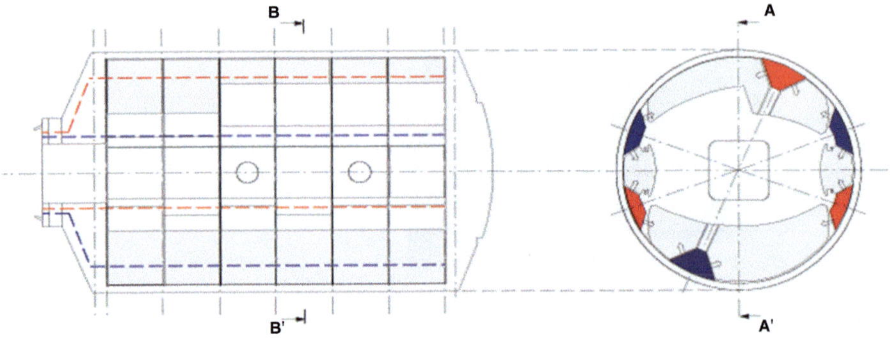

Fig. 2.12 Pinni's asymmetrical design proposal with 6 standoffs & hinged built-in elements

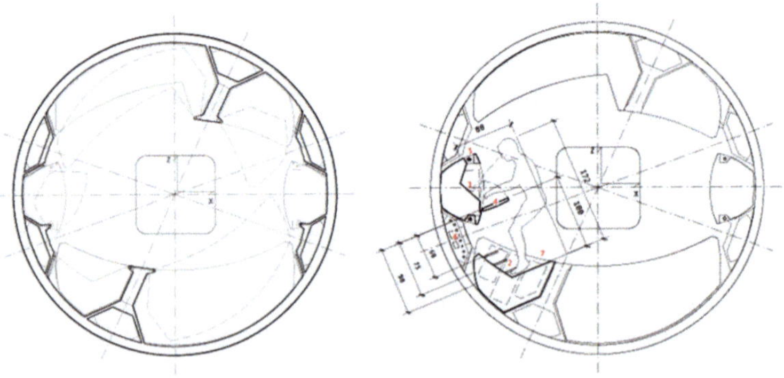

Fig. 2.13 Configuration of hinged elements & ergonomic diagram

should be determined by the optimal amount of space required for its corresponding function or activity, which would help to differentiate them as well as provide more contrast and variety in their experience and use.

Of all the design concepts proposed for habitation in Space, one of the more interesting schemes I have found was one for an ISS module by architect (and astronaut instructor) Martina Pinni (2003) (Figs. 2.12 and 2.13). The design allows the various technical, anthropometric, and ergonomic requirements, rather than the structural framework, to inform the configuration of the interior space. This permits a much more idiosyncratic disposition of elements and spaces unique to its human occupancy as opposed to the imposition of a generic and homogenizing symmetry. The spatial layout avoids orthogonal organization and components, and is much more compatible with the cylindrical geometries of the module's shell. There is an appropriateness in the compartmentalization of the interior by virtue of its functional specificities, and the organic and dynamic forms that result have a greater aesthetic appeal.

Habitats are still classified by their materiality and construction type as well as their engineering complexity, rather than by trip durations or purpose. Space architects at NASA have defined three classes of habitats according to exclusively technical criteria:

Class I: Pre-integrated—components are manufactured and assembled on Earth, outfitted and tested prior to launch. The advantage is that it can be immediately operational upon arrival; the drawback is that due to payload restrictions, the volume and weight of the structure is limited.

Class II: Pre-fabricated—components are manufactured on Earth but assembled when it reaches Space. Partial integration and testing of some subsystems are possible, but outfitting occurs at its destination. The advantage is that larger volumes can be deployed and is less restrictive in terms of launch vehicle size or mass capability; the drawback is that any malfunctioning components cannot be corrected once it has left Earth.

Class III: In-situ—components are manufactured and constructed in the Space venue. All infrastructure, outfitting, integration, and testing occurs on site. The advantage is that there are no restrictions on volume, mass, or launch vehicle size; the drawback is that it requires an on-site manufacturing facility.

In all of the current proposals for planetary sites, the priorities for design revolve around a number of practical issues: efficiency of structure and interior space (e.g., mobile fixtures and flexibility in use), work performance criteria and equipment for work stations, ergonomics, the physical and technical challenges of transport, and of course, safety and survivability of the crew in the Space environment. With the exception of functionality, other considerations for habitability are still rather minimal, continually overshadowed by the engineering demands, as well as by imposed economic constraints. Simple devices, such as the incorporation of color, lighting, texture, greenery, tactile surfaces, and view corridors in the interior can help differentiate, enliven, and personalize what would otherwise be sterile and generic spaces. But engineering is concerned only with physical design problems for which there are quantifiable, empirically verifiable solutions; aesthetic issues are not within their jurisdiction. As a result, they are not taken as seriously as they should be, or dismissed altogether, because they are either misunderstood, or considered to be superfluous. Despite the acknowledgement of more comprehensive habitability criteria in the NASA STD 3000 requirements which offer some sound aesthetic recommendations, translation into the built form still leaves much to be desired.

2.5.3 Speculative Proposals

Alongside the pragmatic approach to habitat design, there are also those who have pursued a more visionary vein, albeit much of it steeped in science fiction consisting of utopian designs that are closer to fantasy than reality. Though they are strong on vision, they are entirely ungrounded in any foreseeable technological capability. But what all of these proposals strive for is to make Space a desirable place to be. Two of the most prominent design proposals in this area are the Dyson Sphere and the Stanford Torus (Figs. 2.14 and 2.15). The Dyson Sphere (and all its subsequent variants) consists of a megastructure framework enclosing a star in order to capture and harness its energy. Unlike the Earth whose outer surface is inhabitable, it is the inner surface of the hollow Dyson Sphere that is intended to be inhabited. Significantly smaller in scale, the Stanford Torus is a large donut structure which rotates about an axis in order to induce artificial gravity. Housed inside the ring are numerous communities and a range of micro-climates with a variety of localized flora.

Somewhere in between the current pragmatism and the futuristic fantasy, is the realm of speculative proposals that are grounded in science but also anticipatory of technological capabilities that have not yet come to fruition. They belong neither in the realm of the fictional nor in the realm of the actual. These latter proposals fall into the category of habitats *intended* for Space, a more ambiguous category than those that are defined as viable on the one end or as implausible at the other. These speculative conceptions, which incorporate both top-down (objective-driven) and bottom-up (capability-driven) approaches, are important because in lying just beyond our grasp—but not too remotely—they can: 1) serve as an operational guide for the development of new technologies, and 2) anticipate and incorporate the newer capabilities that will likely be available in the future, when the goals of space habitation finally come to maturation. This is the realm which holds the greatest potential.

Despite the recognition of the need for a human factors perspective, and the body of research that already exists, 60 years after humans first ventured beyond Earth, Space architecture still focuses almost exclusively on pedantic engineering challenges, such as internal pressurization in the vacuum of Space, dust mitigation on the Moon and Mars, reducing payload to conserve fuel for lift-off, etc. Relatively

Fig. 2.14 Dyson sphere

synonym scant attention has been paid to the quality of the constructed living environment or the well-being of its occupants, as it is considered a lower priority to economic drivers. In his most recent publication, *Social Foundations for Human Space Exploration*, Jim Dator critically points out: "It is curious that even now most actual designs, such as that of the ISS, are still extraordinarily cramped and stark. . .The reason may be said to be economic necessity, but that seems a lame excuse. . .Could it be that we are spiritually and aesthetically impoverished, valuing machines and a kind of 'economics' over humans?" (Dator, 2012, pp. 45–46). Part of the reason why the approach to habitat design has been so pedestrian has to do with the legacy of Space exploration. Traveling to Space was only for the tough and rugged, for those who could face its dangers with a cool, masculinized stoicism. This attitude continues to infuse the design approach today; pioneering Space is meant to be more a gritty adventure rather than a pleasant excursion. As Dator advises, we need to present living in Space in a more favorable light if we are not only to entice humans to migrate to Space, but also to make it more livable once we are there.

Fig. 2.15 Stanford torus

2.6 Deficiencies of the Habitational Approach

Despite the more recent attention on habitation criteria, the conditions for habitability do not seem to be apparent in the current generation of habitat prototypes. They still appear to be driven almost entirely by engineering constraints, and are more reminiscent of the 'Spam-in-a-can' model. The interiors themselves may meet the minimum requirements for functionality and even provide some nominal amenities to offset the starkness of the lifestyle, but from the images available, it is obvious that little or no emphasis has been placed on developing the interiors, nor how occupants use and perceive their space (with the noted exception of Transhab).

Designs being produced in the private sector attempt to make visiting Space more appealing, but tend to revel in banal pseudo-luxury—what one might expect in a quasi-futuristic airport terminal on Mars (Fig. 2.16). In catering to public tastes, the imagery is intended to be reminiscent of familiar earthly locales that speak nothing about the character or experience of their extraterrestrial location.

However, even if the most generous standards for comfort and satisfaction are met, the current habitability criteria still do not take into account other factors which are critical for human longevity in Space. A definition of habitability is insufficient if the design of the constructed environment is approached as a purely external condition that is objective and independent of human perception. A more comprehensive understanding recognizes that habitability is an interactive and

Fig. 2.16 Space habitat interior by B. Versteeg (courtesy of www.spacehabs.com)

subjective condition, that living environments are not just physical spaces with objects in them, but are places that have a temporal aspect by virtual of human presence, and are interpreted and experienced differently by each individual. Without giving the experiential aspects a high priority in design strategies, any habitat is destined to failure from a human perspective.

I would therefore argue that a fourth level should be added to NASA's three existing levels of habitability; one that addresses how humans interpret their surroundings and ascribe meaning to them. The experience of beauty, a connection to our surroundings, a sense of belonging, an environment that allows its occupants to feel fulfilled and purposeful. . .these are human needs of a higher magnitude. This expanded perspective implies a standard that goes beyond the benchmarks of habitability; it speaks to how we *adapt* to our environments. The design features we should be cultivating in the context of adaptability revolve around qualities of resilience, robustness, and endurance—in short, what is required not just for comfort and well-being, but for longevity once we have migrated beyond the protective atmosphere of Earth. Conditions for adaptability, and the critical role that perception and experience play in that process, is discussed in the next chapter.

References

Bannova, O., Landschulz, A., & Smith, I. F. C. (2005). Autonomous architecture: Summit station in Greenland design proposal as a test-bed for future planetary exploration. *SAE Technical Paper No. 2005-01-2909*. Warrendale, PA: SAE International.

Blair, S. M. (1991). The Antarctic experience. In A. Harrison, Y. Clearwater, & C. McKay (Eds.), *From Antarctica to outer space* (pp. 57–64). New York, NY: Springer.

Binsted, K., Kobrick, R. L., O'Griofa, M., Bishop, S., & Lapierre, J. (2010). Human factors research as part of a Mars exploration analogue mission on Devon Island. *Planetary and Space Science, 58*(7–8), 994–1006.

Bishop, S. L. (2006). Psychological and psychosocial health and well-being at pole station. In C. Cockell (Ed.), *Project Boreas: A station for the Martian geographic North Pole* (pp. 160–171). Liverpool, England: British Interplanetary Society.

Bishop, S. L., Kobrick, R., Battler, M., & Binsted, K. (2010). FMARS 2007: Stress and coping in an arctic Mars simulation. *Acta Astronautica, 66*(9), 1353–1367.

Cagle, Y. (2009). *Space adaptation syndrome*. Presentation given at NASA University of Hawaii Astrobiology Institute, Honolulu.
Dator, J. (2012). *Social foundations of human space exploration*. New York, NY: Springer.
Dudley-Rowley, M., Okushi, J., Gangale, T., Flores, P., & Diaz, E. (September 2003). *Design implications of latent challenges to the long-duration space mission*. AIAA Space 2003 Conference (Vol. 24) in Long Beach, CA. Reston, VA: American Institute of Aeronautics & Astronautics.
Evans, G., Stokols, D., & Carrere, S. (1988). *Human adaptation to isolated and confined environments: Preliminary findings of a seven month Antarctic winter-over human factors study*. NASA Contract NAG2-387, NASA Contractor Report 177499.
Gallagher, W. (2001). *The power of place*. Cambridge, MA: Harvard University Press.
Hall, T. W. (1994). *The architecture of artificial-gravity environments for long-duration space habitation*. Doctoral dissertation, University of Michigan, USA.
Hall, T. W. (2004). *Architectural design to promote human adaptation to artificial gravity*. NASA white paper.
Harrison, A. A. (2002). *Spacefaring: The human dimension*. Berkeley, CA: University of California Press.
Harrison, A. A. (2010). Humanizing outer space: Architecture, habitability, and behavioral health. *Acta Astronautica, 66*(5–6), 890–896.
Harrison, A., Clearwater, Y., & McKay, C. (Eds.). (1991). *From Antarctica to outer space*. New York, NY: Springer.
Ihle, E. C., Ritsher, J. B., & Kanas, N. (2006). Positive psychological outcomes of spaceflight: An empirical study. *Aviation, Space, and Environmental Medicine, 77*(2), 93–101.
Imhof, B. (2003). *Musings towards a new genre in space architecture: Transcripts of an architectural journey*. Vienna: Kunst Bundeskanzleramt.
Kanas, N. A., Salnitskiy, V. P., Ritsher, J. B., Gushin, V. I., Weiss, D. S., Saylor, S. A., & Marmar, C. R. (2006). Human interactions in space: ISS vs. Shuttle/Mir. *Acta Astronautica, 59*(1–5), 413–419. doi:10.1016/j.actaastro.2006.02.007.
Kanas, N., & Manzey, D. (2008). *Space psychology and psychiatry*. El Segundo, CA: Microcosm Press/Springer.
Kantermann, T., & Roenneberg, T. (2009). Is light-at-night a health risk factor or a health risk predictor? *Chronobiology International, 26*(6), 1069–1074.
Kaplan, S. (1988). Perception and landscape: Conceptions and misconceptions. In J. Nasar (Ed.), *Environmental aesthetics: Theory, research, and application* (pp. 45–55). Cambridge: Cambridge University Press.
Kennedy, K. J. (2002, October). The vernacular of space architecture (AIAA 2002–6102). In *1st space architecture symposium* (*SAS 2002*), Houston, Texas (pp. 10–11).
Kloog, I., Stevens, R., Haim, A., & Portnov, B. (2010). Nighttime light level co-distributes with breast cancer incidence worldwide. *Cancer Causes and Control, 21*(12), 2059–2068.
Maruyama, M. (1975). *Cultures beyond the earth*. New York, NY: Vintage Books.
Maruyama, M. (1984). Aesthetics and the environment in outer space, subterranean and underwater communities. *Futures, 16*(2), 148–162.
Pass, J. (2006a). Astrosociology as the missing perspective. *Astropolitics, 4*(1), 85–99.
Pass, J. (2006b). The astrosociology of space colonies; or the social construction of societies in space. *AIP Conference Proceedings, 813*(1), 1153–1161.
Pinni, M. (2003). *Interior design for a space module*. Reston, VA: American Institute of Aeronautics and Astronautics.
Poynter, J. (2006). *The human experiment: Two years and twenty minutes inside biosphere 2*. New York, NY: Thunder Mouth.
Preiser, W. (1991). Environmental design cybernetics: A relativistic conceptual framework for the design of space stations and settlements. In A. Harrison, Y. Clearwater, & C. McKay (Eds.), *From Antarctica to outer space* (pp. 147–160). New York, NY: Springer.

Seedhouse, E. (2009). *Martian outpost: The challenges of establishing a human settlement on Mars*. Chichester, UK: Praxis.

Stevens, R. (2009). Light-at-night, circadian disruption and breast cancer: assessment of existing evidence. *International Journal of Epidemiology, 38*(4), 963–970.

Stuster, J. (1996). *Bold endeavors: Lessons from polar and space exploration*. Annapolis, MD: Naval Institute Press.

Suedfeld, P. (1991). Groups in isolation and confinement: Environments and experiences. In A. Harrison, Y. Clearwater, & C. McKay (Eds.), *From Antarctica to outer space: Life in isolation and confinement* (pp. 135–146). New York, NY: Springer.

Swift, D. W. (2006, September). The Voyager spacecraft project – Reasons to remember the grand tour. *AIAA Technical Conference*, 2006–7463.

Vogler, A., & Jorgensen, J. (2004, May). Windows to the world; doors to space: A reflection on the psychology and anthropology of space architecture. In *Space: Science, technology and the arts (7th Workshop on Space and the Arts),* Noordwijk, The Netherlands: ESA/ESTEC (pp. 18–21).

Wapniak, A. A., Dudley-Flores, M., & Gangale, T. (September 2007). *An astrosociological approach to defining indigenous Martian architecture*. AIAA Space 2007 Conference & Exposition in Long Beach, CA. Reston, VA: American Institute of Aeronautics & Astronautics.

White, F. (1998). *The overview effect: Space exploration and human evolution*. Reston, VA: American Institute of Aeronautics and Astronautics.

Chapter 3
Shifting from Habitation to Adaptation in Space

3.1 Premise

Habitability is often equated with adaptation but while habitability is the first requirement for adaptation, a fuller understanding of adaptation encompasses more than issues of physical and social well-being and comfort; it entails evolutional criteria that promote longevity—traits such as resilience, robustness, and durability. Whereas conditions for habitability are generally passive, criteria for adaptability cannot be assessed or cultivated without some kind of engagement with the environment. Well-being and comfort requirements for habitability can be achieved through avoidance of external conditions and without any active participation on the part of the subject, but evolutional traits required for adaptation call for confrontations and negotiations with environmental resistances. Genuine adaptation does not occur in circumstances void of stress or demands. It is through resistant encounters and the overcoming of obstacles that organisms grow hardier and more capable; thus they are more likely to endure.

Adaptation to the physical environment is usually examined through a biological lens, looking exclusively at physiological processes. But psychical aspects also play a critical role in how we adapt to the places we inhabit. How the environment is perceived can prove to be a more important factor in the process of adaptation than the actual conditions of the environment itself. To illustrate this point, one can live in a fortress and still feel insecure and vulnerable, while another can live in an exposed shelter, yet feel safe and secure. This disparity suggests that perception is not necessarily linked to the physical conditions of the environment, but the way in which they are mediated. A focus in habitat design limited to only the physical aspects of the surroundings neglects to take into account how those aspects are interpreted by its occupants and informed by the habitat technologies. If we are to successfully adapt to the conditions of Space, recognizing what our surroundings mean to us will be one of the most important components to assess.

To understand how humans experience their surroundings, a phenomenological perspective must be adopted. Rather than framing an understanding of space as

E.S. Lockard, *Human Migration to Space*, Springer Theses,
DOI 10.1007/978-3-319-05930-3_3, © Springer International Publishing Switzerland 2014

TIMELINE:	past phase (1970's—1980's)	present phase (1990's—2010's)	immediate future (2020's—2030's)	near future (2050's+)
OBJECTIVE:	short-term missions	long-duration voyages	first Mars migration	permanent outposts
DESIGN CRITERIA:	survival	performance	habitation	adaptation
SCOPE OF TECHNOLOGIES:	safety (physical)	capabilities (functional)	well-being (comfort)	resilience (negotiation)

Fig. 3.1 Extended timeline of design objectives in human Space exploration

something objective, phenomenology in the architectural discourse espouses the subjective point of view in which values, history, and culture have bearing on how we perceive and interpret spatial phenomena. Knowledge of the physical world is grounded in experience, not in information. It comes through the various sensory apparatus of the body, and thus is more visceral than rational. Empirical realities are filtered through consciousness. Unlike the objective rationalist approach, conscious presence imbues the physical world with history, culture, symbolism; these are the elements by which we ascribe meaning to the places we inhabit, and also inform our relationships to our environments.

Incorporating a phenomenological approach will be instrumental in guiding the development of technological interfaces towards promoting favorable interactions with the planetary landscape. Driven by the present goals for manned Space exploration, technological intervention occurs at three levels: at the level for survival (as was the case for the earliest manned missions); at the level for functionality (to aid in work performance and productivity); and at the level for comfort and health (which is currently the habitability focus of human factors research). The first level utilizes such interventions as airlocks, pressurization systems, temperature and air quality controls, etc. The second level addresses aspects such as adequate task lighting, sufficient space for activity, appropriate work surfaces, tools and equipment, etc. The third involves leisure activities, intimate communications, food preparation, personal and private spaces, exercise equipment, hygiene, and so on. A fourth level however, must be introduced, aimed at promoting the fourth criterion for adaptation proposed in the previous chapter: technologies of engagement (Fig. 3.1). These must take into account not only issues of habitability, but also evolutional issues, as well as incorporate an experiential perspective that acknowledges the essential role of perception in the adaptive process. The fourth level would explicitly differ from the previous ones in that their purpose would be to offer opportunities to encounter and negotiate resistances in the indigenous Space landscape. Unlike the imperatives for survival, work, or comfort—which generally rely on steering clear of the dangers posed by the outside—adaptation technologies actively establish relationships by fostering interaction with the environment. (Specific technologies will be discussed in Chapter 4.)

3.2 Re-defining Habitability to Include Adaptation: An Evolutional Perspective

Devoid of any consideration of far-future implications, the current approach to designing for long-term habitation refers to a timeframe that can be described at best as 'beyond the short term.' While this research is certainly a critical foundation for human habitation in Space, it is still too myopic in scope. As a result, human factors research on habitability has been constricted to a view of adaptation that is equated more with providing amenities for creature comforts than augmenting somatic capacities in order to be better suited to the extraterrestrial environs. Comfort and well-being are essential for a lengthy stay in Space, but adaptation implies a timeframe beyond the first generation inhabitants as well as a wider range of issues beyond quality-of-life considerations. If we are to insure survival of the human *race*—and not just human *individuals*—in Space, adaptation needs to be thought of in broader evolutionary terms. Human adaptation not only entails comfort and well-being in one's surroundings, but also the ability to endure. It is concerned with both the functional and somatic processes (the physiological component), but also a deeper phenomenological relationship to the physical environment (the psychological component). In the broadest terms it is understood as "any kind of biological reaction which reduces environmental stress and/or increases tolerance against the stressor; it may take the form of population-specific genetic characteristics, physiological acclimation, and learned behaviors. Inherent to these processes is a notion of environment [that] is not limited to natural conditions alone, but embraces the cultural, social, political, and economic reality of humans" (Schutkowski 2006, p. 11).

As the scope of human factors has expanded from survival criteria to include aspects that promote comfort and well-being, there must also be a shift in designing for adaptation that has been thus far derived from limited habitation criteria to also encompass evolutionary considerations. Both shifts imply a qualitatively different focus, but the requirements for 'evolutional' adaptation will be more problematic to address as they may often be at odds with the optimal conditions for habitation (Fig. 3.2). In fact, focusing exclusively on the conditions to insure survival and promote well-being may possibly *impede* evolutional adaptation to the Space environment. Each objective entails a different purpose for technological implementations: for example, technologies to insure *survival* in Space include shielding against cosmic radiation; technologies that address *habitability* will offer leisure activities and ranges of social interaction; and technologies to facilitate *adaptation* must focus on aligning the somatic capacities of the organism with the physical conditions of the environment, as well as focus on phenomenological aspects such as developing a familiarity with, and sense of belonging to, the Space environment. Reconciling these three agenda is highly problematic; even a minimal degree of physical exposure to the Space environment may increase risk and vulnerability of the crew to unacceptable levels. Exceeding the minimums for risk exposure for reasons other than operational necessities runs counter to the imperatives to insure personal safety. However, if we are to adapt (in the fullest

Timeframe: OBJECTIVE →	Short-term: SURVIVAL	Mid-term: HABITABILITY	Long-term: ADAPTATION
Gravity	non-critical	close to 1g	local gravity conditions
Air composition	should cause no deleterious effects	similar to Earth composition	should cause no deleterious effects
Oxygen	above 15%	optimal 21%	eliminate need
Atmospheric pressure	partial pressurization	close to 1 atm	local pressure conditions
Temperature range	tolerable for life	comfortable for life within somatic capacities	ability to withstand extreme climates
Axis rotation (day)	non-critical	close to Earth's diurnal cycle	new planetary cycle
Sun rotation (year)	non-critical	seasons	new planetary cycle
Sunlight	non-critical	similar to Earth exposure	local sunlight conditions
Solar flares (SPE)	shielding	emergency shelter	tolerable exposures to flares
Cosmic radiation (GCR)	underground or regolith shielding	water or electro-magnetic shielding	tolerable exposures to radiation
Scope of factors	physiological	physiological/ psychosocial	evolutional
Timeframe	present (and past)	near future (in our lifetime)	distant future (future generations)
Timescale	months	years	decades, centuries

Fig. 3.2 Criteria for somatic adjustments based on different objectives

sense of the word) to Space environments, our technologies cannot be used to avoid confrontations with environmental resistances.

The prescriptions for psychological and physiological adjustment therefore depend on clarifying what our ultimate objectives are for venturing into Space. Survival alone, because it entails designing a controlled environment to sustain the minimum standards for life, requires in some cases the least physiological adjustment and the greatest technological intervention (where conditions are life-threatening)—but ironically in other aspects, the most physiological adjustment and less technological intervention (under non-life-threatening conditions). Psychosocial needs, for the most part, are irrelevant in the survival scenario. Of the three objectives, habitation requires the least physiological effort on the part of the human because comfort criteria are intended to fall well within our somatic range. Unlike survival or habitation however, adaptation—understood in evolutional terms—demands a more careful analysis of the relationship between the physical environment and our somatic limits. Quantification of those conditions that are conducive to adaptation becomes even more challenging, and we should expect that some criteria for adaptation will diverge from those of either survival or habitation.

The next section delves into different modes of adaptation: the conventional physiological and psychosocial processes, but also introducing a phenomenological dimension.

3.3 Traditional Adaptive Processes

3.3.1 Physiological Adaptation (Biological)

Adaptation is defined in the field of ecology as "a measure of physiological fitness of the organism with respect to one or all of the conditions of its environment and a pliable capacity permitting a system to become modified in response to disturbance. Adaptation is displayed by the adequacy of an organism to cope with the conditions of its natural environment and to utilize its resources so as to maintain its ecological position. It is achieved through an adjustment of its requirements and tolerances to the elements of the [natural] habitat" (Steiner 2002, pp. 34–35). For all organisms there are somatic limits on adjustment capacities; "required to adapt too far from their origins, most institutions will die" (Kelly 1994, p. 381); therefore, introduction to new environments requires finding a balance between the degree of exposure and the capacities of the adjustment mechanisms available to the organism.

Physiological adjustment occurs at four levels, varying from each other in degrees of permanence: regulatory, acclimatory, developmental, and genetic (Moran 2000). Regulatory adjustments are the most flexible and ephemeral, as well as the most frequently occurring, requiring minimal physical commitment from an organism; the body's thermostat is an example of a regulatory adjustment in which we sweat to cool our bodies down when it's hot or shiver to generate metabolic heat when it's cold. Acclimatory adjustments are longer lasting and occur typically after the formative years of development; they are reversible in that when a stimulus that induced change is no longer present the body usually reverts to its former condition. Muscle enlargement resulting from sustained manual labor, or increased lung capacity due to high altitude exposure are examples of acclimatory adjustment. Developmental adjustment is irreversible change that occurs during the formative period of an organism, and thus becomes permanent in the individual. Examples of this are bone mass and height that are influenced by early diet and exercise. The final level, genetic alteration introduced by random mutation, is permanent in the collective population, but usually takes several generations to become widespread and affect a critical mass. Unlike the other three levels of adjustment which are reactions to shifting environmental conditions, genetic change is not caused by, or is a response to, the environment, as was Lamarck's assertion. Genes don't mutate in order to become more favorable to its physical context; the organisms that possess the more advantageous genes are simply more likely to survive and reproduce. Strictly speaking then, genetic change is not an adjustment per se, but rather an outcome.

Unlike other organisms which can only rely on the slow process of genetic change, adjustments in human beings are not limited only to biological processes. "The human body is able to adapt by genetic, physiological, behavioral, and cultural means. The purpose of these various levels of adjustments is to enhance adaptability through a flexible hierarchy of responses" (Moran 2000, p. 9); the more options an organism has at its disposal to react to its vacillating environment, the greater its chances of survival and longevity. Compared to more resilient organisms like bacteria,[1] the somatic capacity of humans is extremely limited. But the cultural and technological means we have developed allow our species to endure in the most unfavorable environments; without these interventions, we humans would have quickly perished outside a narrow range of ecological conditions. Non-biological modes of adjustment, such as the use of technology, generally fall into the regulatory category. Clothing and central heating are examples of regulatory technologies. Though most technologies tend to mediate at the regulatory level, they also affect acclimatory and developmental capacities by reducing the demands on our bodies. The more intensive the technological mediation, the less somatic adjustment is required on the part of the organism.

As we begin to embark on the era of first-generation habitation in Space, technology will serve to make the necessary adjustments within the contained environment so that it is suitable for human occupation. On a physiological level, the burdens borne by the human body will be kept to a minimum; thus the adjustments will be only on a regulatory or acclimatory level. However, if we allow a controlled degree of exposure to select idiosyncrasies of the Space environment, we can reasonably expect significant developmental change with the second generation in Space. "Developmental adjustments. . .provide a more flexible adjustment to prevalent conditions than does genetic change. . .a more rapid mechanism for improving survival chances and enhancing reproduction than genetic change accumulated over several generations" (Moran 2000, p. 8). The most obvious example is the physiological development of the human body under conditions of reduced gravity. Bone density and mass will decrease, and the body may grow more accustomed to equalized fluid distribution rather than concentration in the lower extremities that is caused by the pull of gravity. In conditions of microgravity, vestibular and sensory-motor responses will develop in the absence of a universal vertical orientation. The question of which environmental conditions are acceptable to increase exposure to, and to what degree, is an area that would require further study. Reduced gravity conditions are not immediately life-threatening—though long-term exposure has serious debilitating and potentially lethal effects. But can the human organism be adapted to greater exposure to cosmic radiation?[2] Also to consider is that the capacities for subsequent generations will undoubtedly increase substantially, given their opportunity for developmental adjustment that will not be an option for first generation immigrants.

Which aspects of the Space environment are subject to regulatory, acclimatory, or developmental adjustment depends upon the conditions of the Space venue to be settled. Taking a comparison between environmental conditions on Earth and on

[1] Bacteria can assimilate entire sets of DNA within a 24-hour period (Margulis).

[2] Contrary to lay opinion, cosmic radiation is not an entirely undesirable environmental factor that should be eradicated. Exposure to infinitesimal amounts here on Earth is the cause for many advantageous genetic mutations, which in the long run contributes to diversity, and without which there would be no evolutionary change (Chaisson 2001, pp. 152–153).

	EARTH	MARS
Gravity	1 g	0.3 g
Air composition	78% N_2, 21% O_2, 0.1% Ar	95% CO_2, 3% N_2, 0.1% O_2
Oxygen	21%	0.1%
Atmospheric pressure	1 atm	0.1 atm
Temperature range	-89°C to 58°C	-125°C to 16°C
Axis rotation (day)	24 h 0 m	24 h 37 m
Rotation around sun (year)	365 days	687 days
Sunlight	100%	43%
Cosmic radiation exposure	3.6 mSv per year	250 mSv per year

Fig. 3.3 Comparison of environmental conditions between Earth and Mars

Mars for example (Fig. 3.3), four conditions—air composition, atmospheric pressure, temperature, and radiation levels—are critical to immediate survival, and therefore non-negotiable (at least in the immediate future). In other words, these four conditions will absolutely require some degree of technological intervention to meet an acceptable standard. Those conditions which are critical to survival will be under the exclusive purview of regulatory technological adjustment. Other Earth/Mars conditions that either have great disparity but are non-critical (such as amount of gravity or sunlight), or similar enough to be negligible (length of day)[3] may require—though not necessarily—some intervention, but the nature and the degree is much more negotiable, allowing for variable somatic participation as well; these adjustments fall under the realm of either regulatory or acclimatory adjustment.

As humans change in response to both their natural and their constructed environments over time—and likewise, as the environments they occupy also change through their presence—so will the processes of adaptation. Adjustments in the 3rd, or the 9th, or the 20th year of life in Space can be expected to be of a different nature and magnitude than those of the 1st year, just as those of the second or third generation will be significantly different from those of the first generation. Both conflicting and evolving criteria will therefore, present a technological design challenge in trying to resolve these differences, both in what the priorities appropriate to that time should be, as well as how to anticipate the changes of future generations.

[3] Almost all life that has evolved on Earth has developed circadian rhythms based on a 24-hour light/dark cycle. While organisms can adjust to slight changes in those cycles, the absence altogether of environmental cues that regulate those cycles can cause severe disruptions and physiological dysfunction. However, because those cues change seasonally on Earth, organisms can adapt to some degree of deviation from the 24-hour cycle.

3.3.2 Psychosocial Adaptation (Salutogenesis)

Early analog research on the psychological states of crew members in remote and extreme environments focused exclusively on the negative effects of living under stressful conditions, where behavioral health was evaluated from a perspective of pathology. Much of what has been written examines the worst-case scenarios (see Sect. 2.3.2), and completely overlooks positive accounts. But recent studies which have dispensed with the assumptions of pathology have found that "contrary to prevailing beliefs, the effects of confinement, monotony, reduced stimulation, and solitude are frequently beneficial" (Suedfeld 1997, p. 165). In fact, there is abundant evidence that shows "a decrease in perceived stress or distress as crew members adapt to the environment" (Ritsher et al. 2007, p. 337). In analog studies "Antarctic winterers not only coped with stressors, but actually became stronger and healthier," recounting their experiences as characterized by camaraderie, insightfulness, and increased sensitivity. Researchers concluded that the "challenge also brings out fortitude, perseverance, ingenuity, and comradeship. Many polar workers experience a strong emotional bond with the place" (Suedfeld 1997, p. 166).

These positive effects resulting from adverse conditions are known as 'salutogenesis'—a phenomenon whereby "powerful experiences enhance or bring about well-being and personal growth" (Kanas and Manzey 2008, p. 135). This approach is very different from previous assumptions. "Whereas older notions of psychological well-being rested on the absence of pathology, behavioral health implies positive, rewarding interactions with both the physical and social environments" (Harrison 2010, p. 891). Despite the overwhelming list of potential harmful effects attributed to ICEs, they have also proven to induce positive outcomes, such as increased self-awareness, satisfaction in overcoming adversity, development of strong social bonds, a sense of mastery and accomplishment, appreciation of the grandeur of Space, and pride in being part of an elite team (White 1998; Harrison 2010). Contrary to initial expectations, respondents in studies demonstrated a "surprising ability to endure, recover from, overcome, and even be strengthened by, events that to outside observers seems overwhelmingly destructive" (Suedfeld 1997, p. 164). It has since been recognized that "missions in space…can provide mental health benefits, even though the experiences may be fraught with danger, hard work, and isolation from family and friends" (Ritsher et al. 2007, p. 339).

The Overview Effect was one such example that demonstrated the salutogenetic effects of seeing the Earth from afar. There are numerous accounts from astronauts who claim after seeing their home planet from Space that they experienced a oneness with all humanity and with the Earth itself; many even describe having feelings of transcendence and increased spirituality, and upon returning back to Earth have a greater appreciation for its beauty and wonder (White 1998; Harrison 2010).

Psychological studies in Space exploration have historically followed the traditional approach, focusing solely on the range of possible deleterious reactions to the multitude of stressors in Space. This bias informed selection of crew members, favoring those that demonstrate the ability to stoically withstand the anticipated hardships of long durations away from Earth and loved ones, prolonged confinement to a constrictive habitat, and constant exposure to high degrees of risk and danger. Crew-vetting concentrated on finding astronauts with the "right stuff"—those who could not only cope but also perform well under extreme duress. But Suedfeld advises a shift from seeking invulnerability to developing resilience, from pathogenesis to salutogenesis, from the isolated individual to the integrated group (2005).

Surprisingly, the impact of the design of the habitat on the mental states of its occupants was never considered. Rather than searching for astronauts who can tolerate highly uncomfortable living conditions, why not redesign those conditions to ameliorate environmental, psychological and social stressors? Or better yet, why not design the habitats to promote greater salutogenetic effects? Design is based on optimization of conditions within given constraints; it should not only focus on mitigating the negative effects of ICEs, but also facilitating the positive ones. "Maximizing crew members' chances of salutogenesis is one way to offset their risk of pathogenesis" (Ritsher et al. 2007, p. 336). Environmental psychologists, as well as architects and interior designers, understand well the impact of the built environment on people and can offer the technical knowledge to improve them, but unfortunately this is a domain of expertise that continues to go untapped. A leading researcher in the field of salutogenesis, Peter Suedfeld reports that despite soliciting expertise in this area, "the response of the NASA participants to a long list of suggestions and recommendations for research from the social scientists was dismissive" (Suedfeld 2005, p. B61). Integrating them into the design decision-making process and giving credence to their proposals would be the first step.

Despite the current efforts to make Space habitats more comfortable, I would argue that conditions of adversity are in fact *necessary* for salutogenesis to occur—that these beneficial, health-enhancing effects can only be experienced through moderated exposure to risky, dangerous, and unpleasant circumstances. Sense of purpose and accomplishment, cooperation, forming of close bonds and attachment to place, intimate knowledge of and relationship to the environment can only come through confronting challenges and overcoming obstacles. Behavioral health does not imply the *absence* of interactions with both the physical and social environments; there must be some kind of engagement. The task is to design for those interactions to become positive, beneficial ones. However, "people can deal with these environments only because the adversity is not extreme" (Suedfeld 1997, p. 166); that is, adversity alone is not sufficient. It must be balanced with some degree of comfort as well.

The experience of adverse conditions is the necessary basis for adaptation to new environments. We mustn't attempt to eliminate them entirely through technological interventions. If we create habitats that are aimed to minimize or even eliminate exposure to environmental stressors, we effectively design away the opportunities for salutogenesis to occur.

It would be shortsighted to characterize salutogenesis as merely an incidental consequence of enduring hardship; we should consider how it can actively be elicited through technological mediation with our environment. On face value this antithetical proposition may seem absurd and even reckless. I am not proposing that we do not attempt to carefully mitigate the risks and dangers of living in Space; but that we don't do away with the intense *experience* of those stressors.

Adaptation follows from developing resilience. "Space travelers have demonstrated a remarkable capacity to endure and thrive—in short, to adapt—under the stressful circumstances of space missions, some of very long duration" (Ihle et al. 2006, p. 93). This speaks to the need for encountering resistances and challenges as part of our experience of the environment. "Adaptive functioning results from a complex interaction of external environmental pressures and personal qualities. When external pressures are extreme (as in Space or in life-threatening situations) the motivation for adaptation is strong" (Ihle et al. 2006, p. 94). Design considerations and definitions of habitability have to move beyond the scope of achieving comfort towards cultivating robustness. Among other things, "adaptation to the mission is important, because if crew members are feeling better, they will be more likely to complete their tasks accurately and safely" (Ritsher et al. 2007, p. 337). Though greater efficiency and minimization of human error is vital, the importance of adaptation should not be reduced to assessments of work performance. If the human race is to endure in Space, we need to frame adaptation in larger evolutional terms.

3.4 Incorporating a Phenomenological Perspective

The process of adaptation is rooted in how we respond to our environments not only physiologically but also cognitively; that is, how we perceive the physical phenomena around us. In fact, subjective experience of the environment is more critical than the objective characteristics itself (Suedfeld 1991; Dator 2012). This experiential component is not under the purview of the engineer or technician, who are concerned only with functionality and performance, and not with emotive, aesthetic, or interpretive matters.

Take a window, for example: an engineer would design a window to meet quantifiable physical criteria: interior and exterior pressure differential, structural reinforcement to span the opening, expansion and contraction coefficients of the glass, etc. An architect on the other hand, would design the window with an understanding of the subjective component: the impact of natural light and exterior views on mood and behavior, the sense of security afforded by visual awareness of one's surroundings, the feelings of vulnerability of being viewed from the outside, and so on. The window is not simply a transparent surface or opening in an otherwise solid wall. It also represents a penetration of barrier, a threshold between domains of interiority and exteriority, and a cognitive portal to a world beyond. The window may be a successful feat of engineering and yet still fail miserably from the user's perspective if these phenomenological aspects are not taken into account. Understanding how we develop meaningful relationships to our surroundings, and how these principles can be applied to extraterrestrial environments, is therefore essential if we are to successfully adapt to life in Space.

The phenomenological aspect of dwelling in Space—that is, how Space will be *lived and experienced*—is as essential to habitability and long-term adaptation as is the physiological. But what our environments *mean* to us—as opposed to how we *function* in them—has been for the most part, ignored in habitat design.[4] Our unconscious assumptions, attitudes, and beliefs about the Nature implicitly inform how we approach the design of technologies that mediate our relationship to the environment. The prevailing Western attitude, grounded in both Darwinian and rationalist ideologies, is that Space is a hostile and inhospitable place; this preconception is reflected in the language of conquest that is often used in conjunction with Space exploration. Given this mindset, technological design has tended to focus exclusively on protecting human explorers from its dangers, rather than exposing it to its wonders.

If we are to mitigate the sense of isolation and confinement that will most certainly be an aspect of life in the remote environments of Space, this approach must be revisited, lest we reinforce the perception of Space as 'foreboding' or 'perilous' through the design of the habitat interface. Dator advises: "Not-Earth should be portrayed as 'Hawaii,' not as a hostile wasteland where all must live as prisoners. But that is not the kind of analogs we have studied...we need to stop calling space 'hostile' or 'an extreme environment'" (Dator 2012, p. 46). If we continue to describe Space in these terms, we will only perpetuate an antagonistic relationship with the extraterrestrial environment through our technologies, one in which the Space environment will be treated, at worst, as an obstacle to be overcome, and at best, as something to be exploited in service to humankind's needs.

It is not only the Space environment that is viewed as hostile, but the presence of humans as well. Astrobiologists are focused on how to minimize contamination, both of humans and of the planetary environment—particularly on Mars where there is an increasingly strong possibility that life may exist. Though exercising foresight and caution is advisable, this approach implicitly treats human presence as a pathogenic condition. We see ourselves as hostile interlopers in a pristine, unmarred, planetary landscape; instead, humans should view themselves as a coalitional presence in the Universe. We should augment our scientific efforts to also better understand how our presence is beneficial to the new environments we are exploring, and how we can create symbioses between humans and new environmental entities. "An individual and its environment [are] best understood not as separate entities but as a dynamic feedback system" (Capra; Lovelock; Gallagher, 2001, p.103).

We need to move away from depicting Space as an inhospitable place, and start employing technological interfaces not only to protect human life but also to acknowledge the *beauty* of Space. Rather than concentrating only on the physical

[4] With the exception of a few, such as Angel Marie Seguin, Andreas Vogler and Jesper Jorgenson.

challenges presented by the Space environment, we should also recognize how the environment *enables* us to inhabit it. By re-framing our approach to living in Space, we can finally cease trying to 'conquer' it, and learn how to 'co-habitate' instead.

3.4.1 Making the Foreign Familiar

> Imagine if you can, a mass of canyons, ravines, ridges, gullies, chasms, and mountains, piled one above another in inextricable confusion...no vegetation anywhere to be seen, nothing but stones around us...this black yawning abyss just before us...The gloom increases with every step. The walls assume in the darkness a thousand grotesque and misshapen forms. The obstacles in our pathway become more frequent and dangerous; the darkness more and more intense...We go on, hesitating, doubting, fearing, until after hours of tedious toil such as I hope never again to experience, we finally reach the bed of the river that has worn this mighty wrinkle in the face of Mother Earth.

This is an excerpt from nineteenth-century pioneer Samuel Woodworth Cozzen's journal, dramatically titled "Journey of Death," documenting his impressions upon discovering the Grand Canyon while venturing through the southwest. One might expect a similar reaction by the first pioneers on Mars, if they had never seen or expected such a landscape before. It is only natural that any new encounter with that which is completely foreign or unknown will be intrinsically alienating at first. Even landscapes that today we find awe-inspiring were probably at one time perceived as threatening and treacherous.

The Grand Canyon today has come to be one of the most spectacular and wondrous geological formations on the planet, drawing visitors from all over the world. There are several factors for such a radical shift in perception from feelings of terror to feelings of awe: one factor is the degree of potential danger posed, either actual or imagined. A person may perceive a place as more threatening when they themselves feel more vulnerable, despite whether there is a real danger or not. A second factor is the duration of exposure to the new environment. As Dator notes, "people's initial attitudes towards an environment typically change with prolonged exposure to it" (Dator 2012, p. 46). All places, independent of the level of danger, inevitably become more familiar and less threatening with time spent there. And thirdly, how the environment is perceptually mediated through interventions is another critical variable; here is where architectural technologies can play a vital role in influencing our attitudes towards our environment. If the habitat is designed in such a way to serve exclusively as a protective shelter—in other words, to shield the occupants as much as possible from its environment—it constitutes a phenomenological negation of the existence of its surroundings. When the wider surroundings of one's immediate perceptual field are not acknowledged, a cognitive detachment results; this in turn exacerbates both feelings of loneliness and the perceptual tunneling of the external world that accompany prolonged isolation.

In any planned outposts on the Moon or Mars, cultivating a cognitive connection to the lunar/Martian landscape should therefore be a requirement of the architectural program inasmuch as creating comfortable private and social zones or efficient work spaces. Acquisition of data about the environment, such as the task that the Mars Rover performs, is not sufficient alone for promoting cognitive attachment; because "thought creates distance and destroys the immediacy of direct experience" (Tuan 1977, p. 148), the process must also include an experiential component beyond mere information retrieval. How is the environment experienced through the senses? How do we interpret beauty and meaning in our environments? How does it make us feel? Take, for example, how we develop a sense of security: it can be cultivated through total isolation from dangers of the environment (actual) or by developing trust through a process of familiarization with the surroundings (perceptual/experiential). But each entails differing design strategies. The former, in looking at the environment as objective information, leads to an architecture of hostility by shutting out the environment, whereas the latter, in appealing to the subjective meaning of that information, leads to an architecture of negotiation through engagement with the environment. (Moreover, a structure that is relatively secure can still be perceived as unsafe—demonstrating that there is no necessary correlation between perception and empirical reality.)

3.4.2 Empathy and Alienation

According to art historian Wilhelm Worringer, the world that each of us synthesizes for ourselves is derived from our tendencies towards both empathy and the abstraction, reflecting the degree of psychical connection or detachment we feel towards our empirical world. Empathy is the capacity to which we can connect to an entity external to ourselves. We have empathy for things that are familiar to us, that resonate psychically within us. Empathy is a condition of perceptual interiority or inclusion. Conversely we tend towards feelings of apprehension when initially faced with things which are foreign and unknown to us. Worringer writes that the urge to abstraction in earlier societies reflected an "immense spiritual dread of space" and "a psychic attitude towards the cosmos" (Worringer 1997, p. 13). Our abstractive capacity seeks the representational, not the real, and space—by providing the third dimension—brings the phenomena of our perception back into the realm of the real. Moreover, he posits that the degree to which humans feel alienated in their surroundings, the more they strive towards an aesthetic abstraction of them. The act of abstraction is, therefore, our response to give order to, or to make sense of, alien/alienating external phenomena through aesthetic manipulation; hence, abstraction reflects a condition of perceptual exteriority or exclusion. Whereas empathy is the act of remembering, abstraction is the act of imagination.

Empathy and abstraction are complementary drives. When the conditions that give rise to empathetic connections are lacking, we become nostalgic and melancholy for the past and things that are familiar to us. On the other hand, when the conditions that demand the use of our imagination to satisfy our abstractive urges are lacking, then we become disengaged from the world around us, and our realm of perception becomes increasingly insular and solipsistic. "Just as the desire for empathy as the basis for aesthetic experience finds satisfaction in organic beauty, so the desire for abstraction finds its beauty in the life-renouncing inorganic" (Worringer 1997, p. 4). Taken together, empathy and abstraction are reciprocal extensions of our need to ascribe meaning to our world. They are as such aesthetic expressions, for we find beauty both in what we already know and covet, as well as in what we can envision and conjure.

This tension between the need for the mundane and familiar, and the yearning for the strange and exotic is even reflected in our visions of utopia: "Humans have always searched for utopias…an essential similarity found in all of them synthesizes "two antipodal images: the garden of innocence and the cosmos" (Hillis 1999, p. 179). This is illustrated by the observation that astronauts aboard the ISS gaze at the Earth for the very opposite reasons for why we gaze at the stars. While stargazers dream about the unknown Universe, astronauts were dreaming about their home planet. Similarly, studies show that crews at Antarctic research stations created virtual windows or put up images of idyllic places that they have either visited in the past (evoking nostalgia) or would like to visit in the future (evoking fantasy) (Carrere et al. 1991). Yi-Fu Tuan writes: "Place is security, space is freedom; we are attached to one and long for the other" (Tuan 1977, p. 3). Outer Space—the 'great outdoors'—is the ultimate metaphor for both chaos and freedom, because it is by definition the realm of the undiscovered and unknown. We need this experience of alienation because without it we would cease to imagine, just as if we were to realize the house of our dreams, we would cease to dream (Harries 1998). We need dreams of the past (the subject of empathy and memory) as well as dreams of the future (the subject of abstraction and imagination).

The focus on well-being and comfort in habitat design can overemphasize the need for empathy and familiarity, and overlook the need for abstraction. Feelings of alienation towards new environments, to a degree, appeal to our urge to abstraction; it is the a priori response to the realm of the unknown and the foreign. This urge to abstraction provides us with the essential challenge required for adaptation and growth; it is the *process* of overcoming obstacles through the facility of imagination. Alienation, a form of psychical resistance, is the wellspring for challenge and creativity; confrontation with the unknown begins from a state of alienation, but experience—a combination of time, interaction, and ascription of meaning—is the path to empathy. Integrative technologies that create a physical and psychical interface between us and our environments are also technologies that promote the transition from alienation to empathy.

As with the exploration of any new territory, the introduction of new technologies is also alienating at first, but as we grow accustomed to them, they become unconscious extensions of ourselves, so seamlessly ingrained in our daily lives that we hardly take notice of them anymore. McLuhan recounts that "the effect of electric technology had at first been anxiety. Now it appears to create boredom. We have been through the three stages of alarm, resistance, and exhaustion that occur in every disease or stress of life, whether individual or collective" (McLuhan and Lapham 1964, p. 26). The process of developing empathy through experience, along with the processes of somatic adjustment discussed in the previous section, are both at the core of adaptation; therefore in addition to performance criteria, technological mediations should be assessed as to whether they are conducive to this meta-effect of fostering engagement with the extraterrestrial environment, rather than suppressing it.

Architecture responds to our empathetic and abstractive needs on the most visceral level through establishing domains of interiority and exteriority in our living environments. Each passage through a door is a transformative one that requires a cognitive shift to adapt to different spatial conditions (threshold of alienation). Each gaze through a window is an act of assimilation in the world, a cognitive portal that connects us to people, objects, and places outside of ourselves (generator for empathy). These are some of the elements that can provide those essential psychical connections to the outside world (Vogler and Jorgensen 2004). Living in confined environments can sever our relationship to the external world. Without a relationship to the 'outdoors,' a confined environment will never be fully experienced as a place of dwelling but rather as a place of incarceration. In order to develop an empathetic connection to extraterrestrial environments, we need to find novel ways in which to experience them. Without this experience an introversion occurs; the context by which we reference our thoughts and actions shrinks, thus increasing the tendency to perceive only our immediately surroundings as real. Because Space lies wholly outside the realm of collective human experience, it relates far more to our abstractive capacity than to our empathetic one; hence, generating empathetic connections should be given careful attention in the design of our living habitats in Space.

As we eventually establish settlements and venture out even further into the solar system, our definition of outer Space will change accordingly. Lagrange points, the Moon and Mars—in becoming gradually more familiar to us through sensory experience—will no longer be considered "outer" Space, but rather as places within our perceptual and experiential proximity, which in turn will push the perceptual boundaries of outer Space further beyond. The 'new' outer Space will then be referenced in relation to inhabited loci in the inner realm of Space, much in the same way we reference the Moon and Mars in terms of their distance and size to Earth.

3.4.3 Hermeneutics of Space

How the concept of 'Space' is translated in different languages, and the metaphors associated with them, say much about the degree of affinity or alienation we feel towards Space. Dator observes that differing attitudes are reflected even in the monikers that have been assigned to spacefaring vehicles: while NASA names their spacecraft after fearsome Greek or Roman gods, like Mercury, Apollo, Gemini, Saturn, Jupiter, Orion, and Titan JAXA, on the other hand, references folklore or elements of Nature, such as Kibo (hope), Kizuna (wind), Kiku (chrysanthemum),

Fig. 3.4 The Void

Hinode (sunrise), and Kodama (spirit) (Dator 2012, p. 6). Culturally ingrained attitudes profoundly influence how technologies are applied as well as the disposition to adapt to new and strange environments. Views towards Space can be described as one of five metaphors (Lockard 2005):

3.4.3.1 Space as Non-world

A word for 'Space' in Chinese is 'tiān kōng,' which literally translates as 'empty sky', signifying that Space is a container, devoid of any presence within it (Fig. 3.4). In Vietnamese, the word 'không gian' similarly equates Space with nothingness. These metaphors are rooted in feelings of alienation, reflecting both trepidation and anxiety towards migration to Space, which represents the abyss, the void, a state of 'nowhereness.'

3.4.3.2 Space as the New World

Americans, whose history and cultural spirit are founded on pioneerism, tend to conceive of Space as something to be explored and conquered (Fig. 3.5). The use of the term 'outer Space' in the English language clearly indicates that Space is viewed as an external entity, distinct and remote from ourselves and our cognitive sphere. This metaphor is also derived from feelings of alienation, but in this case also reflect attitudes of conquest and imperialism.

3.4.3.3 Space as the Free World

The title of Tsiolkovsky's treatise, "Free Space," reveals this Russian's attitude towards Space as benevolent, in which micro-gravity is seen as a liberating

Fig. 3.5 The Wild West

Fig. 3.6 M.C. Escher's "Liberation"

condition (Fig. 3.6). Space is not at all a hostile or threatening prospect; on the contrary, it is where humanity would not only evolve to insure its own survival, but ascend to greater happiness. Derived from the tendency towards abstraction, Tsiolkovsky's image of Space reflects emancipation and freedom from the constraints of Earth.

Fig. 3.7 Japanese tea room

Fig. 3.8 Heaven

3.4.3.4 Space as the Same World

The Japanese have a very different attitude towards Space; they see themselves as an integral part of the Universe, not as distinct from it. Their term for Space—'uchuu'—views the cosmos as if from the inside rather than from the outside, as a constituent rather than as an observer. Traditional Japanese architecture strives for harmony and integration in its natural surroundings. The Japanese teahouse (Fig. 3.7) serves as a microcosmic representation of the Universe, in which the architecture and the ritual of the tea ceremony symbolize the realm of Nature and human interaction within it. Unlike the previous cultural metaphors, the tearoom is derived from the disposition for empathy, reflecting feelings of interrelatedness and connectedness with the cosmos.

3.4.3.5 Space as Meta-World

In religious and spiritual contexts, Space and the Universe represent a metaphysical existence, possessing an other-worldly or ethereal quality (Fig. 3.8). The Christian notions of Heaven and Hell directly relate to perceptions of Space and gravity: 'the starry heavens above' (in the sky, free of gravity) and 'firey hell below' (at the core, the point of maximum gravity). Also derived from the tendency towards abstraction of the unknown, metaphysical metaphors likewise reflect a state of transcendence beyond the confines of Earth.

3.4.4 The 'Geography' of Space

> We continually personalize and humanize the given environment as a way of both adapting to it and creating order and significance.—Mihaly Csikszentmihalyi

Geography refers to "the science that studies the lands, the features, the inhabitants, and the phenomena of the Earth." The field deals with spatial analysis of natural and human phenomena, area studies (places and regions), and the study of the relationship between humans and land. Although the Greek root 'geo' refers specifically to the Earth, a more generic interpretation would refer to any planet in general. Prospective human habitation on Mars calls for an investigation of these issues.

To be able to dwell somewhere means, first, to know where one is, and second, to be familiar with one's surroundings. Dwelling implies that one feel a sense of territory and an intimate knowledge of their living environment; therefore dwelling cannot occur without a sense of place and context. Geographical discourse seeks to understand how we define our places and how they acquire social significance. Developing a 'geography of Space'—one that is unique in it idiosyncrasies, but built upon the traditional geographies of Earth—is equally important in the effort to promote adaptation as the physiological processes that are more commonly associated with it. The most fundamental question is: how can a sense of place be established in Space?

There are three modes of determining place: absolute location, relative location, and phenomenal location. *Absolute location* can be pinpointed using quantitative devices such as grids, quadrants, topography, and axes; distances can be precisely measured but are not described in terms of relations to other locations or from the human perspective. *Relative location* can be established contextually—independent of knowledge or experience of a particular place—by describing it with respect to other known places, and referring to landmarks or monuments rather than relying on quantitative descriptors. And finally there is *phenomenal location*, one that is established through presence, either through direct experience or through history. Phenomenal location can be a physical place, such as the neighborhood one grew up in, or a virtual place, such as an internet chat room.

These various methods of mapping are not merely about fixing location, but also establishing relationships between subject and environs, both proximal and remote. The same behavior can be observed in migrating animals, who likewise create their own cognitive maps of their habitat by inspecting the peripheral territory and establishing graduated boundaries of interiority and exteriority. The inner most sanctum is where intimate social activities take place, such as eating, sleeping, and raising the young; the larger sphere—their operational domain—is where the search for food, protection of the habitat, and other routine activities occur. "Spatial structures and forms in 'geographical' space are often used to provide context for interactions" (Dodge and Kitchin 2001, p. 57). Although the other modes of mapping location in Space are worth expounding upon, the focus of this section is on developing phenomenal location that is essential to adaptation.

3.4.4.1 The Concept of Place in Space

The notion of place is a human construct; it is the *act* of dwelling that transforms space into place. If space—in defining the realm beyond what we have intimate knowledge of—is an abstract concept, then place is what we know through experience and as loci of control (de Certeau 1984). Places are *lived* spaces. They have not only specific location, but also history and meaning imbued by human presence (Entrikin 1991). Place conveys a sense of finite interiority; space, on the other hand, represents infinite exteriority. We cannot know space except in the abstract; space has neither location, nor scale, nor reference. It is in fact, placelessness (Casey 1996). Whereas Space is where movement occurs, place is destination, a "pause in movement" (Tuan 1977, p. 138).

Human presence, however ephemeral, is the first requisite in constructing place in Space. The televised scene of Neil Armstrong's first steps on the Moon and the planting of the American flag holds immense historical meaning to us, and it took only this symbolic gesture to transform the Moon into a place (Fig. 3.9). That single historical moment, albeit brief, was enough to change the Moon to *lived* space. Prior to the Apollo landing and the first words uttered from its surface, the Moon was merely an object, a *conceived* space (LeFebvre 1991). That event irrevocably altered our collective consciousness: Space had gone from being *uninhabitable* to merely being *uninhabited*. With repeated expeditions to the Moon, what was once foreign to us will gradually become ordinary. The initial Moon missions shifted our boundaries of perceived interiority and exteriority in the Universe. Reconnaissance technology, as an extension of our senses, has now to some small degree expanded our sphere of interiority to the edges of the solar system, and as we continue to explore our surrounding galaxy, that sphere will continue to expand.

Because place refers to a specific locality which has presence and history, places resonate with our empathetic tendencies. It contextualizes by establishing phenomenological relations of 'here' and 'there.' More importantly, places instill a sense of belonging. "The concept of *territory* [is derived from] the external physical reality that results from our crafting of place" (Hillis 1999, p. 167). Our proprioceptive orientation, an affirmation of our corporeality, is fundamentally grounded in the understanding of place. Place is synonymous with rootedness, groundedness—terms that are always associated with land. Can a sense of place be evoked in the vast open space of the cosmos, where there is no terra firma? Adaptation is location-

Fig. 3.9 The Moon becomes a 'place' in Space

specific, and without addressing this phenomenological aspect of placeness, Space inhabitants will be unable to develop 'roots' to their new 'homeland.'

The absence of gravity in LEO only exacerbates this rootlessness. Gravity orients bodies by its directional force, and thus provides the datum by which we reference the physical world all around us. Crew members aboard the ISS complained that the absence of vertical cues in the station made using as well as navigating through, the spaces confusing. Orientation became one of the primary issues that informed the interior design of the ISS modules. This issue is less problematic for planetary sites however, as they provide at least some degree of gravity as well as physical landscape.

The construction of scale is also essential for defining place. Scale is not an absolute condition—it is not "simply a 'hierarchically ordered system' placed over pre-existing space." It is a human construct, a "produced societal metric that differentiates space" (Marston and Smith 2001, pp. 615–616), in which affiliations of identity and regimes of social interactions are established. Scalar configurations are the outcome of "the flux of socio-spatial and environmental dynamics" (Swyngedouw 2004, p. 132) that also inform the strata of social relations.

The formation of scale in Space is problematic because we are accustomed to measuring in distances based on Earth's size, which are at most several thousands of miles. But our terrestrial frame of reference is insignificant in comprehending interstellar distances. The distance to the Moon for example, the closest celestial body to the Earth, is approximately 240,000 miles, and once we travel to other planets, those distances are measured in millions of miles at the very least. The distances between interstellar entities are too vast to viscerally apprehend, and therefore become perceptually inaccessible. Developing a sense of scale in Space will be essential in order to introduce layered relationships within the larger surroundings and create hierarchies of territory. It will also be indispensable in establishing relative location to intermediary entities. But perhaps the most profound effect scale will have will be on broadening kinships and our circle of empathy to include the scale of the planet.

3.4.4.2 Home-Making in Space

> People dream of ideal places. The Earth, because of its varying defects, is not viewed everywhere as humankind's final home.—Yi-Fu Tuan (1974, p. 114)

Home is the seminal place from which we create and reference all other places. It is "the base, the center, existentially-speaking, of any world. . .a point of departure. . .Home is the node from which we link to other places and things" (Heim 1998, p. 92). As the source of security and familiarity, it is the origin of our empathetic connection to the world. Whereas the house is a *tectonic* construct, a physical structure with a fixed geographical location, the home is a *phenomenological* construct defined by individual experience and perception of interiority; it need not even be defined by physical geographies. The house is the physical mediator between us and the space beyond it, a negotiation between its structure and the immediate environment, and the home is the psychical one, a site of social negotiation between its occupants and the people outside of it (Massey 1994). We construct the space of home to define our personal territory and control, but also to define our identity through affiliation. Social and cultural spaces beyond the home are also constructed to establish varying degrees of belonging or non-belonging, of inclusion or exclusion.

Social research has acknowledged the importance of recapturing a sense of home in Space (Carrere et al. 1991). This will be even more challenging and crucial as Space—the origin of our alienation—represents everything that home is not: mystery, foreignness, and the imagined/unknown. But thus far, the focus on habitat design in the Space sector has been on building a house equipped with the necessities for survival and practical functioning, and not on creating a home to give a sense of belonging, to interface with the surroundings, and to facilitate adaptation to the local environment.

So then how do we establish home in Space? This is a question that is viewed with skepticism by many. Mander argues that "Our species, *Homo sapiens*, emerged from the chemical soup and soil that is this earth. We are part of an intricate web of life that exists only here. Nowhere else in the universe could possibly be 'home', however ingenious we become in re-creating Earthlike environments in space" (Mander 1991, p. 159). Mander's view exposes his biases and even contradicts his own views. For him and others who object to humans inhabiting Space, Earth is perceived as an entity separate from the rest of the Universe. But according to the principles of deep ecology that he espouses, our planet is as much a part of the larger solar system as each life form is to the larger ecosystems here on Earth. It doesn't occur to him that the only difference between migrating from Earth to Mars and migrating from the tropics to the tundra is only one of scale.

The concept of home can be instantiated at a variety of scales: the family (a kinship conception), the house (a physical conception), the community (a social conception), the country (a geographical conception), or one's place of origin (an historical conception). These variations are expressed through terms such as hometown, homeland, or heimat. Home at its most intimate scale is experienced through the space of the body; at its most encompassing through homeland or heimat. There seems to be a limit to meaningfulness of scale: we generally don't refer to 'home continents' or 'home hemispheres.' Tuan warns that we should be suspicious of topophilia, a love for all land, in the way that we are suspicious of claims to love all humankind (Tuan 1974). When all space can be interiorized then there remains nothing that represents either 'outside' or 'other', eliminating the realm of alienation and abstraction as part of experience. However, this is likely to change as we migrate to Space and view our planet from afar. For human inhabitants on the Moon or Mars, Earth will be referred to as their 'homeplanet', extending the scale of home and adding an anthropological conception to the range of phenomena we call home. Departure from our biosphere will establish a new realm of exteriority, a larger domain of non-homeness by which to re-define the planet itself as home. As the birthplace of humanity, Earth as homeplanet is analogous to the German concept of heimat, which represents the origins of a folk—a region that is defined by genealogy, lineage and history, but not necessarily by particular events or individual experience or memory (Heidegger 2008). For the first time in human history, we have the ability to observe the Earth from beyond, as an outsider would. To be able to visually take in the Earth's boundedness, wholeness, and finiteness, will only serve to strengthen our sense of it as home. Encountering new life forms will likewise reinforce our sense of humanness by introducing a new domain of otherness/outsider.

Although harsh geographical conditions such as what we find on Mars—steep terrain, volatile climate, and infertile soil—may initially induce feelings of alienation, this does not at all preclude the capacity to feel at home in them. In fact, none of the conceptions of home that were described earlier are grounded in the physical conditions of the environment. Like the Moon and Mars, vast regions of the Earth also have conditions that are unfavorable to human life—the depths of the ocean, the high altitude of mountain peaks, the frigid cold of the polar latitudes—but the history of life repeatedly shows that all species have migrated to regions where environmental conditions are extremely unfavorable to their survival, and they have not only survived, but prospered in them, which demonstrates that the psychical aspects of home (e.g. history, culture, genealogy, etc.) can trump the harshness of geography. The weight of lived experience, of memory, is so formidable that not even destruction through natural disaster or war may diminish one's sense of home; a history of hardship and resistance may, in fact, actually reinforce it. If then similar psychical attachments can be forged in outer Space—though it may require generations—then it is certainly possible that humanity may one day also be able to call the Moon or Mars home, despite their unforgiving conditions.

Maintaining a connection to exterior conditions is important because the outside world sets the context for the interior condition we call home. The construction of our personal domain is synthesized in reference to what lies beyond it. Thus because home is subjective and personal, the external world—which lies outside our sphere of control—defines the boundaries of our larger lived reality.

The weaker our experiences are of the outside world, the more disconnected we feel from that larger reality.

3.5 Aesthetic Considerations

One aspect of habitability that has been grossly overlooked in human factors research is aesthetics—a field that is commonly misunderstood and undervalued. It cannot be reduced to mere stylization or attractive packaging—more properly, it is about the experience of beauty and how we ascribe meaning to the world we inhabit. Aesthetic quality should not be dismissed as a non-essential aspect of our normal lives; the perception of beauty pulls us out of the mundane and pedestrian, and allows us to feel a larger sense of purpose and appreciation. For a practical life we need the functional and efficient—but for a fulfilled life, we need the aesthetic.

Beauty can be experienced through Nature (e.g., plants, animals, landscapes, skyscapes) and it can also be experienced through our cultural artifacts—music, literature, works of art, and architecture. Aesthetic appreciation is a critical part of our daily existence. Reading a work of literature, listening to a piece of music, or gazing at the sunset, are all events that make our lives meaningful in ways that functional or comfort criteria cannot satisfy. Even seemingly mundane events such as mealtime provide aesthetic experience: crew members on Space stations often look forward to both the preparation and the consumption of food—for reasons that go well beyond appeasing hunger and enjoying the taste of food: it offers an opportunity for relaxed social interaction, for leisurely activity, and for creative expression, as well as alleviates boredom and under-stimulation of habitat life (Binsted et al. 2010).[5]

Though it is an aspect of human factors that has not been taken very seriously yet in the Space industry, analog research at least acknowledges the importance of aesthetics: "Although the 'pleasingness' of a built environment might not be directly related to crew effectiveness, interior features certainly contribute to overall habitability, and in indirect ways, the presence or absence of particular features probably affects individual adjustment and productivity. . .habitat aesthetics might be an important factor during future expeditions and other remote duty involving long-duration isolation and confinement" (Stuster 1996, p. 189). Numerous studies in environmental psychology confirm that exposure to natural light, exterior views, and the presence of plants in our living spaces boost scholastic and work performance as well as positively affect mood.

[5] Lecture by Kim Binsted at the University of Hawaii School of Architecture, on her experience as a crew member in the FMARS habitat simulation on Devon Island (2007). Studies show that food has been found to help alleviate boredom and monotony in the confinement of the habitat (Suedfeld and Steel 2000), and additional studies are currently underway on the Island of Hawaii to evaluate the impact that the menu has on a potential Mars-bound crew.

Most of the literature on habitation in Space advises that Earth-like conditions should be simulated as much as possible—much like the way first-generation immigrants tend to re-create their original home environments—so as to ameliorate the symptoms of homesickness and isolation that are experienced during extended missions to Space by restoring a degree of familiarity and comfort within the crews' surroundings. However, the side effects of this prescription are that it also inhibits the development of attachment to their new surroundings. As a result, immigrants to Space will suffer from a sense of being perpetually alien—similarly to the immigrants who choose to surround themselves with the trappings of the past and never attempt to adjust to their new culture.

The habitat architecture can facilitate a connection to the exterior conditions, or can likewise sabotage it by severing the crew from the external world entirely. The habitat must establish a precarious balance between both serving to protect the lives and health of the crew and simultaneously engaging them with the very thing they must be protected from. Some degree of engagement is essential because the ability to sensually experience the 'outdoors' on a planetary outpost would not only help to alleviate feelings of confinement, but would be crucial to the adaptation process. Since the unaccessorized human body cannot tolerate the exterior conditions of any non-Earth environment, the technology of the habitat must go beyond its mandates for basic life-support and work performance functions and also be designed to afford a variety of human–environment interfaces.

An example of this type of interface is the viewing portal on the International Space Station. Technical designers were initially reluctant to include windows, as they were considered impractical from an engineering standpoint. But after giving in to the pressure from astronauts and Space architects, windows were finally installed in the newer modules, and they turned out to have a profound effect on the crew: astronauts spent 80% of their leisure time gazing at the Earth through the window (Vogler 2004, p. 4). Astronauts stationed on the ISS also reported experiencing a transcendental state as well as a sense of a greater unity with both all life on Earth as well as with the cosmos, simply by gazing at the Earth from afar, a phenomenon known as the Overview Effect. On a phenomenological level, this portal provided a number of unanticipated positive effects, including: a visual connection to 'home'; a chance for personal reflection; an awareness of a reality beyond their immediate enclosure; a proprioceptive sense of place and location relative to Earth; and a spiritual revelation that everything in the Universe was interconnected. The power of this one simple design feature demonstrates the need to externalize stimuli for psychological well-being. Other features should be employed to do more of the same. They needn't be complex or costly. A number of options are proposed in the following section.

Every site location affords a unique and privileged vantage point that should be taken advantage of. By periodically focusing on the planetary surroundings—both near and distant—cognitive interactions with the exterior landscape are encouraged which help develop empathetic connections. Biologist E. O. Wilson cites our genetic predisposition for preferential views: "There is an evolutionary foundation for the common human preference for sweeping views from an elevated perspective, especially for views involving bodies of water. . .natural selection has predisposed members of the human species to favor views and scenes that in many ways replicate optimal living conditions during the millions of years of human evolution: an unobstructed view of a plain and a nearby pond, lake, or seashore from a safely elevated vantage point" (Stuster 1996, p. 192). Heightened awareness of the external environment also staves off the tendency towards solipsism to which crew members are particularly susceptible in ICEs.

3.6 Design Recommendations to Enhance Experience of the Space Habitat

Given the importance of aesthetic features, I propose a number of possible supplemental strategies that would enrich the experience of both the interior of the habitat as well as the exterior landscape:

3.6.1 Lighting Effects

In addition to creating lighting conditions within the interior of the habitat as zeitgebers for the crew, diverse and changing illumination could occur on the exterior. There are a number of ways this could be done through the introduction of artificial phenomena: a rotating lighting system that mimics Earth's diurnal cycle to indicate motion and passage of time; the simulation of sunrises and sunsets; lightening storms, rainbows, northern lights and other lighting phenomena that occur in Nature on Earth (Fig. 3.10). In addition, there should be some means to highlight the aesthetic features of the extraterrestrial landscape which would draw one's attention outward and attenuate the oppressiveness of incarceration. These 'spontaneous' and 'unpredictable' events would help to alleviate the monotony of life inside the habitat by creating a sense of surprise, mystery and awe, but also enlarging the sphere of perception beyond the confines of the enclosure, by visually engaging the occupants with their external surroundings.

3.6.2 Spatial Contrast and Differentiation

There has been abundant research on ways in which the configuration of the habitat enclosure can be made to feel less constrictive in order to mitigate feelings of confinement. The size of a habitat is limited by the prohibitive cost of transporting larger payloads into Space. However, more important than increasing the size of the habitat to accommodate different demands, is providing contrast within it.

Fig. 3.10 Synthesized external phenomena

Designing for differentiation can go a long way in addressing some of the common problems associated with long missions in cramped quarters (Carrere et al. 1991). Rather than attempting to design for all the different spatial conditions required to support a range of various activities—or on the other extreme, making all internal spaces generically 'multi-use'—the habitat interior should be designed to create spatial dichotomies, so that the *character* of the spaces are qualitatively different from one another, or can be changed significantly. Activities can to some degree be adapted to spaces not intentionally designed for a specific use. This counterintuitive approach allows for experience to be as varied as possible within the limited space of the habitat while still supporting necessary operational demands. Contrasting spaces also give rise to non-programmed activities and different ways of engaging the space, which would help to overcome some of the boredom, confinement, and under-stimulation of daily life on a remote planetary outpost. Instead of spaces programmed specifically (or generically) for *functions* such as lab work, exercise, and leisure, spaces would be designed to elicit varied *experiences* within it. A 'regimented' space for example, might be one that is orthogonally configured to impart a sense of orderliness and efficiency, and would thus support more organized, rigidly prescriptive activities; whereas a 'soothing' space might have soft surfaces and organic forms which would be more conducive to relaxing or unwinding activities. Within each of those spaces a number of different activities or functions—both anticipated and unanticipated—could take place. Research in a systems approach in Space station design supports this approach:

> The systems approach (Miller, 1978) appears appropriate in this area of endeavor—it is linking in holistic fashion diverse phenomena that influence relationships between people and their surroundings, including physical and social environments. . .Differentiation, not universal solutions, in the built environment is being promulgated as a way to solve problems in the future. Such differentiations are based on the recognition that there are different user types, each with special requirements concerning the built environment; that there are different types of places, each with setting-specific requirements; that there are different types of requirements for environments; and that there are different activities that occur in the same environments at different times (Preiser 1991, pp. 149–150).

By creating contrasting interior spaces within the overall enclosure, occupants can enjoy varied experiences in qualitatively different zones. The purpose is not to simply eliminate confining spaces but to allow the occupant to experience a range of qualitatively different spaces which in turn will contribute to greater stimulation. Creating a single large, unmediated, open space does not in fact, lessen the sense of confinement—it just makes for a larger prison; whereas creating a mix of both intimate, closed, contained spaces with *relatively* larger, public, open spaces is more effective, though still not a substitute for the experience of exterior space. Mobile elements that rotate, pivot, slide, fold, collapse, or are removable allow the user to manipulate their interiors accordingly. Private space could then be used either as a workstation, or for quiet contemplation, entertainment, exercise, etc. as well as intimate communal space-making. Not only is this approach practical in terms of economy of space by accommodating different functions and activities, but the experience of variety—expressed through change of scale—is preferable over unbounded space, which is equally disconcerting in its lack of human scale or reference.

Changes in scale give us opportunities to see things anew; "radically different scale of landscapes lifts us out of the rut of habitual perception" (Gallagher 2001, p. 73). In applying this to the design of the interiors, the difference in volume between open public spaces and compartmentalized private spaces within the habitat structure should not be just a matter of degree but relatively pronounced. Reframing spatial contexts also encourages a greater variety of social configurations and interactions.

3.6.3 *Perceptual Dissolution of Barrier*

Confinement is perceived as inviolable boundaries defining inside and outside, so creating a perception of choice as well as contrast *within* the contained environment can mitigate this problem. One way is through the attenuation of thresholds between spaces to create zones of decreasing interiority. The purpose is not to simply eliminate confining spaces but to allow the occupant to experience varied types of spaces which in turn will contribute to stimulation. Degrees of disintegration of exterior barrier through the use of glass curtain walls, ribbon windows, portals, etc., blur the explicit boundary between interior and exterior, and radically alter the experience of enclosure. Transparent rooms are perceived as more permeable and hence less confining, and therefore will offer a refuge to offset the sense of confinement in other areas of the habitat (Fig. 3.11). They don't actually have to be transparent, but offer the *perception* of transparency.

3.6.4 *Qualitative Modulation*

Modulating elements can create adaptable spaces which the crew could transform so as to possess altogether different qualities, and not merely larger or smaller spaces, or accommodation of different functions. These elements could be mobile,

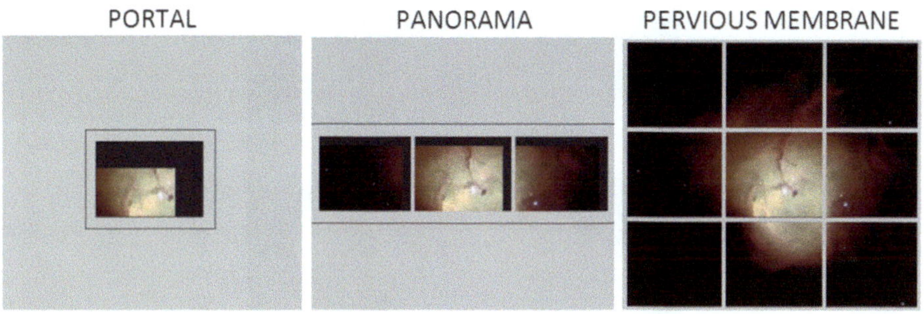

Fig. 3.11 'Dematerialization' of enclosure

constructible, acoustic, visual, textural, etc. This notion of modulation is conceptually very different from genericized, multi-purpose spaces which are designed to have no idiosyncratic character at all, and are thus considered to be flexible. This is an erroneous assumption in architectural design that has resulted in lifeless, bland, uninspired living spaces. Generic spaces may not, in fact, *hinder* a particular activity; however, nor do they *support* any either. In breaking with conventional wisdom, areas within the limited space of the station should *not* be designed to be function-specific unless the operational requirements for a particular activity call for a high degree of specificity. For functions that entail less demanding requirements, areas should be designed to establish a particular atmosphere or character which are then conducive to performing a wider range of activities within it. For example, lab work would likely require a very ordered and well organized area with sufficient storage for equipment. Such a use mandates a dedicated space that adheres strictly to the necessary program for it. However, an area with a soothing, relaxing, or tranquil character lends itself to a range of activities such as reading, napping, massage, writing, meditating, working, yoga, etc. It is not in a strict sense a place for leisure or recreation per se, but rather one that is conducive to many compatible activities; and since it is essentially undefined functionally, allows for new, unintended uses as well. This design approach gives the crew an element of choice and control, which allows them to be active participants in shaping their constructed environment to some degree. This creative, negotiative process would offset some of the boredom, predictability, and under-stimulation that crew members tend to suffer from in restrictive environments. Living spaces and furnishings should of course, be comfortable, but also challenging and adaptive. The more interactive they are, the more they engage the user. They should also be flexible enough that their full potential can be discovered through the imagination of the user.

Fig. 3.12 '1300 Inflatables' and other flexible tensile membranes

3.6.5 Malleable Membranes

The use of inflatables has been proposed as an alternative to solid structures; they are light, compact, and durable, and are able to provide more internal space than a rigid structure for much less the cost. The technology behind inflatables could be further exploited to produce flexible membrane appendages (comprising a hybrid of fixed structure + inflatables) for temporary extensions of the planetary station (Fig. 3.12). (HDU, one of NASA's most recent habitat prototypes, has already incorporated this strategy.) These appendages would function on principles similar to that of a spacesuit: occupants would first unseal the airlock connecting the deflated membrane to the main station in order to pressurize it, after which they could enter the inflated vestibule. These attachments could be temporarily inhabited for personal use, for intimate communal activities, or when there is an occasional need for supplemental space. This too would give crew members an enhanced sense of privacy and of 'getting away from it all.' The option for supplemental space would also help address the condition of confinement by providing dynamic and interactive elements. Whereas a fixed structure rigidly defines the boundaries of inside and outside, malleable enclosures, if they were possible to engineer, would make the perception of these boundaries more ambiguous. For example, if the exterior building envelope were a tensile membrane that could be stretched or extended well beyond its normal perimeter, then what was once exterior space has been annexed to the interior. Moreover, because the enclosure itself can be manipulated, it gives the occupant a greater ability to control their environment. If we take this idea even further and create membranes that were so flexible that one could venture far beyond the initial enclosure, the perception of interior space would almost disappear. Each venture outward would be a uniquely individual one. It would be experientially similar to that of a spacesuit tethered to the mothership, where one could explore space that has not been parceled by structure.

3.6.6 Aesthetic Qualities of Protective Shielding

Water and regolith shielding, which will be necessary for protection against constant cosmic radiation and intermittent solar flares, could potentially be utilized to have aesthetic applications. The use of regolith will probably be the most practical material for shielding, since it is abundant in-situ, weighs less than liquid (putting less stress on the structure), and will not be a precious commodity in Space like water. Regolith can provide adequate shielding against the less intense cosmic radiation, but a small water-encased chamber would be more suitable to provide the increased shielding necessary against episodic solar flare events.[6] Some portion of the chamber should be transparent so that the water could be visible from the interior. The experience of being enveloped in water would also very likely have positive psychological effects on the station's inhabitants (Fig. 3.13).[7]

3.6.7 Boundaries of Personal Territory

Territoriality, according to Csikszentmihalyi, is "one of the most basic needs among animals, including humans" (Csikszentmihalyi and Halton 1981, p. 122). The act of territorializing—an expression of the need for privacy and personal control over one's environment—is therefore inevitable; so the interior should accommodate that impulse by designating compartments dedicated to each individual crew member only. Analog studies in Antarctic research stations showed 'hot-bunking'—shared

[6] Both water and regolith are thermal masses, and thus transmit heat easily. A thermal break between the shielding material and the shell of the enclosure is therefore recommended.

[7] The water in the shielding layer could be used for cleaning or other purposes, and recycled. Or if non-toxic antifreezes are developed, it could even be populated by aquatic species who demonstrate tolerance to cosmic radiation, turning it into a celestial 'aquarium'.

Fig. 3.13 Water shielding

sleeping spaces assigned by work rotations—had negative effects on crew members and should be practiced only temporarily or as a last resort.

3.6.8 Transitional Zones

Another strategy to help relieve the sense of confinement is to re-establish a direct relationship to the exterior by redefining what is perceived as external space. Here we can refer to terrestrial analogs: porches and balconies for example, are transitional spaces that mediate between the house and outdoors (Fig. 3.14). They allow distant and panoramic views of the landscape, sounds and smells of Nature, fresh air, and natural light. They offer an experience of the outdoors, while still enjoying the protections and security of being within the home. Habitats in Space should similarly have some kind of ancillary enclosures that serve the same purpose as porches and balconies do on Earth. By raising awareness of external phenomena, the habitat architecture can encourage extroversion, ward off egocentric tendencies, counteract sensory deprivation, and mitigate the sense of confinement. (EVA's are one way to experience the 'outdoors,' albeit mediated by cumbersome equipment. But this is an act of *leaving* home, which is not the same as bringing the experience of the outdoors *into* the home.)

3.6.9 Non-controlled Conditions

As automated systems will be a requisite part of life in Space, occupants could become physically and psychically disengaged from their internal living environment. It is necessary that positive challenges which facilitate a cognitive engagement with the physical surroundings be incorporated into the living environment. User

Fig. 3.14 The porch

control over the interior environment is essential to a sense of well-being and personalization. Whereas the state of *having control* is a static condition, the process of *controlling* is a dynamic one; the difference between the two is vast in terms of how we interact with our surroundings. For example, the experience of the sea from a cruise ship, where navigation is automated, is very different from that of a sailboat, where the sailor must use the conditions of the wind and the water in order to navigate it. The habitat architecture can address the issue of control by providing spaces that promote appropriate degrees of spontaneity, discovery and playfulness. This can be accomplished through use of color and light, texture and materiality (which is seldom considered), contrasting and dynamic forms, flexible functions—or even elements that can alter themselves, grow or mutate. Variable stimuli would help to stave off boredom and monotony; unpredictable elements would give relief from the tedium associated with long-duration missions and allow the crew members to see things anew. Another negative aspect of the highly-controlled interior is that it can foster complacency and passivity. "Technology [of interiorized living] poses insidious threats to delicate interactions with the natural world that our species has developed over centuries" (Gallagher 2001, p. 205). An example of a self-regulating feature would be a plant-responsive climate control system, where the needs of the plants dictate the interior micro-climates, rather than by a random program or one that is human-controlled. This again would also help humans maintain awareness of the presence and needs of other forms of life outside of their own.

3.6.10 Reactive Environmental Features

Other alternatives for cognitive engagement with the surroundings include the incorporation of features that react to fluctuating weather conditions, such as wind chimes, sails, or other objects that move or respond to the vicissitudes of the planetary climate (Fig. 3.15). Lunar regolith or Mars dust could be sculpted or landscaped. These aesthetic features could be periodically relocated, rotated out of sequence, or altered to rekindle a sense of novelty.

Fig. 3.15 Wind sculpture by Janet Echelman

3.6.11 Presence of Other Life Forms

Anecdotal evidence from Mars simulations and other ICE studies indicates that the presence of plants in the enclosure is also psychologically beneficial—not only useful for food supply and scientific experiments. Plants had a soothing and comforting effect on crew members, and became objects for extended gazing.[8] Research shows that "when the subjects' responses to a stimulating, plant-filled, homey setting and a grim, institutional one were contrasted, the only reaction they all shared was a decline in vigorous activity and increased feeling of fatigue in the austere environment" (Gallagher 2001, p. 17). Plants provide much more than visual distraction however. Because they need to be cared for, the presence of other life forms on board redirects our focus away from our own needs. It brings Nature into our living spaces, it satisfies our aesthetic needs, it reminds us of the symbiotic relationship we have with other forms of life, and unlike the rest of the mechanized habitat, it grows and changes. Plants also naturally improve indoor air quality—one of the most critical factors in interior design for maintaining human health—by removing toxins in the air, optimizing humidity levels, absorbing CO_2 and emitting oxygen (Kopec 2009). In addition to plants, animals provide wonderful companionship (as any pet owner will attest); they reduce stress, lower blood pressure, fulfill the need for physical contact, and alleviate boredom and loneliness (Poynter 2006; Suedfeld 1997; Walton and McConocha 1996) (Fig. 3.16). The presence of companion animals in therapy research shows that they also help immeasurably in overcoming hardship and adversity.

[8] Lecture by Kim Binsted at the University of Hawaii School of Architecture, on her experience as a crew member in the FMARS habitat simulation on Devon Island (2007).

Fig. 3.16 Companionship of other life forms

3.6.12 Organic Elements

Humans respond more empathetically to other organic entities than to inanimate objects. We have a need for some connection to Nature, which is why we often bring plants and animals into our living spaces. In environments where there is a deficit of living things, humans yearn for this attachment. Because mechanistic environments are non-organic in their nature, they constitute another form of alienation. Inflatable structures on the other hand—in contrast to the coldness and sterility of mechanistic structures—are very appealing (Figs. 3.17 and 3.18). Besides the practical advantages inflatable structures offer as viable habitats in Space, they may also prove to be the most aesthetically satisfying living environments as well. Inflatable structures have a natural similarity in their forms to other organisms, something which is not present in the mechanistic structures. Their surfaces are soft and pliable. They "come to life" when they are inflated and invite tactile engagement. There is even something playful about them. It's possible that we respond to them in such a visceral way because they recall an empathetic connection to our primordial home, namely that of the womb—our first encounter with an 'inflatable structure.' The use of water shields to protect against cosmic radiation would also reinforce a womb-like character in the enclosure.

These various approaches not only maximize utility by creating efficient, multi-functional spaces, but on a psychical level they also offer variety, stimulation, control, and choice to the user. The user is therefore no longer a passive, reactive participant in their space, but an active one who has the power to reconfigure their dwelling spaces to accommodate different activities as well as a range of social settings from the communal to the private.

3.6.13 Virtual Scenery

Research shows that people respond positively to images of Nature. Representations of natural settings evoke feelings of comfort; they are in a sense, virtual environments because we experience the imagistic representation through the sense of sight. Virtual realities that are experienced additionally through the other senses

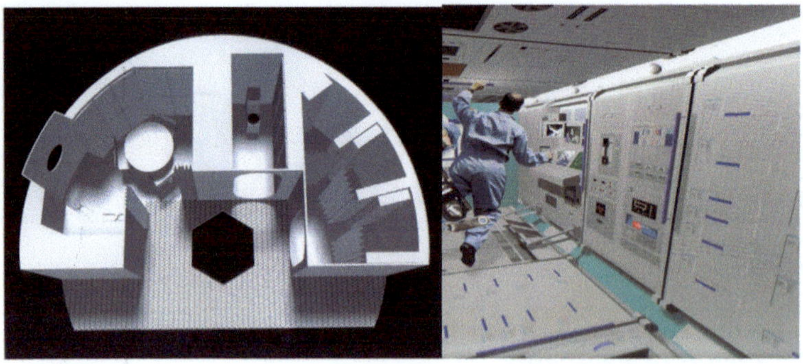

Fig. 3.17 Artificial forms, mechanistic interiors

Fig. 3.18 Organic form and materials

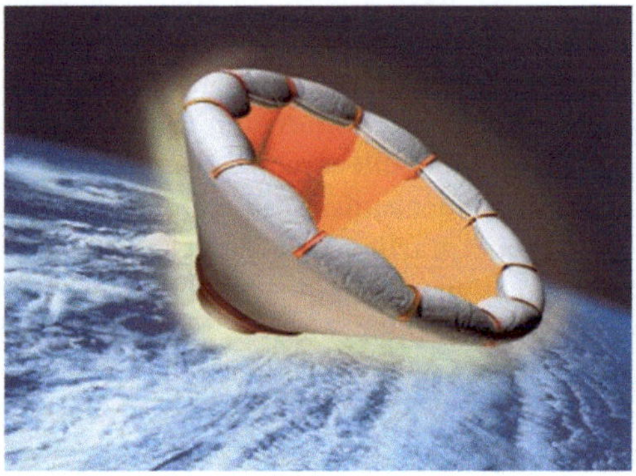

become even more engaging. Instead of merely gazing at a picture of the ocean, imagine if we could not only see the movement of the water, but also hear the waves crashing on the shore, feel the mist of salt water on the skin, and smell the moisture in the air—the experience of the simulation would be far more powerful. Virtual scenery could afford opportunities for experience through simulation of things that are not available in Space. A virtual fireplace for example—the flickering of the flames, the crackling of the burning wood, the smell of smoke, and the feel of warmth—could make the interior space of the habitat more cozy and intimate, which would be reinforced by the knowledge of the subfreezing temperatures beyond the confines of the protective enclosure (Fig. 3.19). The development of such virtual technologies aimed at offering a more comprehensive sensory experience would not require sophisticated devices.

Fig. 3.19 Virtual hearth

3.7 Evolutional Factors: The Need for Resistance and Negotiation

The most fundamental way our bodies are engaged with our environments is through the presence of resistance; "resistant materiality [marks] the experience of living as embodied creatures" (Hayles 1999, pp. 15–16). Resistant forces develop an organism's resilience by creating disequilibria to which the organism must then respond; "we face material limitations around which we must negotiate, compromise, and move. It is by doing so, however, that such constraints contain the seeds of opportunities" (Hillis 1999, p. 203). These confrontations are vital to the process of adaptation, without which, living things would ultimately cease to evolve. All life on Earth has evolved in reaction to its particular resistances. The planet's gravitational force, for example, gave rise to the unique morphologies and structures of living things, as well as to the geological and climatic conditions of the planet. And without the presence of gravity, the Earth would not have been able to retain the atmosphere which made it a favorable place for life to appear and flourish in the first place.

It is also through engagement and negotiation with the vacillations of the physical world that we come to know and fully dwell in the environments we inhabit. As Tuan puts it: "experience is the overcoming of perils" (Tuan 1977, p. 9). We respond to obstacles through our capacity for abstraction—our urge to establish order in chaos, affinity in hostility. This instinctual capacity is the wellspring of all creative endeavor. But as with empathy and alienation, a balance must also be struck between conditions of resistance and receptivity, as both are complementary aspects of lived experience. Because life can be either extinguished or perpetuated by resistant encounters, resistances must be tempered, though not eliminated. In excess they can result in annihilation, but in right measure, they can initiate growth and change.

We are predisposed to think of resistances in our lives—be they physical (e.g., walking up a steep hill or enduring cold climates), environmental (e.g., pest infestations), intellectual (e.g., learning a new language), or sociological (e.g. arguments or social conflicts)—as something disagreeable or burdensome, and seldom as something to be relished or embraced. Dealing with resistance is inefficient, consuming both time and effort. Our tendency is often to avoid or eliminate them, and not surprisingly, many of our technologies have been invented to serve this purpose.

But if we attempt to maximize creature comforts and minimize resistances in our built environments, we create a state of passive inhabitation and consequently lose the ability to interact meaningfully or constructively with our environment. In our quest for developing novel technologies intended to improve our quality of living, we can misguidedly strive to eradicate resistance in exchange for greater ease, convenience, and security. As a result, technology can "foster an estrangement from this world, which flows from technology's utility in 'pushing back' an often hostile natural world that for millennia was perceived as limiting human intentionality" (Hillis 1999, p. 202). Technology effectively becomes a form of anesthetization, dulling our senses and inducing complacency. Without some degree of resistance in our daily lives, we would become passive and complacent. It is through embracing resistance—an engaging, negotiative process—that we find authenticity of experience; avoidance through technological intervention serves to diminish that authenticity.

Once humans travel to Space, they will be introduced to different kinds of resistances from what they are accustomed to on Earth. Astronauts who have spent months in the micro-gravity environment of the International Space Station suffer from extreme muscular atrophy and bone decalcification; as a result, when they return to Earth they have lost most of their physical strength. We can expect the same problems when we travel to Mars, which has a third of Earth's gravity, and the Moon, which has only one-sixth. While weightlessness might be experienced as a liberation from the gravitational constraints of our home planet (at least this was Tsiolkovsky's romantic view) it also deprives us of the challenge to physically overcome counterforces—the constant negotiations with our environment which keep us robust. Given the extremely deleterious effects of long-term exposure to micro-gravity on the body, artificial gravity via centrifugation—a technology of resistance—will need to be introduced. The purpose for creating resistance in this case however, has more to do with avoiding infirmity than for the sake of the experience of gravity itself.

On the surface it may seem an odd proposition to argue that technologies should preserve or infuse resistance in the living environment. But the adaptation process for those living in Space may be impeded if technological mediation does not permit sufficient exposure to resistant conditions. The dilemma however, is that the technologies required to support life will be of such a high degree of intervention that they may effectively inhibit the essential process of negotiation and adaptation. Issues involving life safety cannot be compromised, but other conditions that are not essential to our immediate survival, such as micro- or partial gravity, can be phased gradually. Though the absence of gravity over the long term is detrimental to human health, it is not directly or immediately life-threatening, and so the intervention of a compensatory technology (e.g. artificial gravity environments through centrifugation) could be routinely calibrated to introduce a process of long-term adaptation. Eventually emigrants to Mars or the Moon should become acclimatized to reduced gravitational forces, so that they can fully adapt to their new living environment.

The experience of resistance should not only be introduced sporadically; it should be incorporated into the daily living routine. "The process of struggling with the forces of nature powerfully focuses the mind...We're relieved of petty concerns and the monotony of routine, feel an immediate sense of purpose and the value of life" (Gallagher 2001, pp. 75–76). Our constructed habitats, in serving our needs, should not deprive us of necessary resistant encounters. According to biologist Lynn Margulis, "physical contact is a non-negotiable requisite for many differing kinds of life" (Margulis 1998, p. 5). In service to the larger goal of long-term adaptation, the habitat should avoid focusing exclusively on shielding its occupants from the dangers of their environment and instead develop various ways of engaging and interacting with it.

In summary, current habitability criteria are insufficient for long-term adaptation and should be supplemented by the inclusion of an experiential component. In addition to their functional requirements, habitat technologies must assist fragile human aliens in developing resilience, robustness, and endurance for Space environments to which they are not accustomed, by promoting physiological acclimation to conditions of the new environment. They must also establish conditions that give rise to salutogenesis (encouraging psychological health rather than treating psychological pathology). Aesthetic experience will be integral to inculcating meaning and purpose in the lives of the astronauts. We must not overlook the importance of developing a sense of belonging, by establishing place and home in the Martian landscape; this cognitive connection will lead to greater empathy and familiarity with the surroundings.

Adaptation does not rely only on the biological vehicle; humans exploit other means at their disposal. "Like no other species, humans have emancipated themselves from the genetic fixations and reaction norms of their basic biological equipment; and they have developed and extended their adaptability by non-genetically coded means...By means of specific strategies of environmental use, humans found successful solutions as responses to stresses and limiting factors in their habitats, which enabled then to settle in areas quite remote from their places of origin" (Schutkowski 2006, p. 251). The next chapter will examine a range of different options for technological interfaces with the non-Earth environment, and then determine how each of these might promote long-term adaptation.

References

Binsted, K., Kobrick, R. L., O'Griofa, M., Bishop, S., & Lapierre, J. (2010). Human factors research as part of a Mars exploration analogue mission on Devon Island. *Planetary and Space Science, 58*(7–8), 994–1006.

Carrere, S., Evans, G. W., & Stokols, D. (1991). Winter-over stress: Physiological and psychological adaptation to an Antarctic isolated and confined environment. In A. Harrison,

Y. Clearwater, & C. McKay (Eds.), *From Antarctica to outer space* (pp. 229–237). New York, NY: Springer.
Casey, E. S. (1996). How to get from space to place in a fairly short stretch of time: Phenomenological prolegomena. In S. Feld & K. H. Basso (Eds.), *Senses of place* (pp. 13–51). Santa Fe, NM: School of American Research.
Chaisson, E. J. (2001). *Cosmic evolution: The rise of complexity in nature*. Cambridge, MA: Harvard University Press.
Csikszentmihalyi, M., & Halton, E. (1981). *The meaning of things: Domestic symbols and the self*. Cambridge, England: Cambridge University Press.
Dator, J. (2012). *Social foundations of human space exploration*. New York, NY: Springer.
de Certeau, M. (1984). *The practice of everyday life*. Berkeley, CA: University of California Press.
Dodge, M., & Kitchin, R. (2001). *Mapping cyberspace*. London: Routledge.
Entrikin, J. N. (1991). The betweenness of place. In J. N. Entrikin (Ed.), *The betweenness of place: Towards a geography of modernity* (pp. 6–28). Baltimore, MD: Johns Hopkins University Press.
Harries, K. (1998). *The ethical function of architecture*. Cambridge, MA: The MIT Press.
Harrison, A. A. (2010). Humanizing outer space: Architecture, habitability, and behavioral health. *Acta Astronautica, 66*(5–6), 890–896.
Hayles, N. K. (1999). *How we became posthuman: Virtual bodies in cybernetics, literature, and informatics* (1st ed.). Chicago, IL: University of Chicago Press.
Heidegger, M. (2008). *Basic writings*. New York, NY: Harper Perennial Modern Classics.
Heim, M. (1998). *Virtual realism*. Oxford, England: Oxford University Press.
Hillis, K. (1999). *Digital sensations: Space, identity, and embodiment in virtual reality* (1st ed.). Minneapolis, MN: University of Minnesota Press.
Ihle, E. C., Ritsher, J. B., & Kanas, N. (2006). Positive psychological outcomes of spaceflight: An empirical study. *Aviation, Space, and Environmental Medicine, 77*(2), 93–101.
Kanas, N., & Manzey, D. (2008). *Space psychology and psychiatry*. El Segundo, CA: Microcosm Press.
Kelly, K. (1994). *Out of control: The new biology of machines, social systems, and the economic world*. New York, NY: Basic Books.
Kopec, D. (2009). *Health, sustainability, and the built environment*. New York, NY: Fairchild Books.
LeFebvre, H. (1991). *The production of space*. Oxford, England: Blackwell.
Lockard, E. S. (2005). *Design considerations for long-term habitation in space: The role of aesthetics and human cognition*. Presented at the 10th Annual International Symposium on Space Exploration, International Space University, Strasbourg, France.
Mander, J. (1991). *In the absence of the sacred: The failure of technology and the survival of the Indian nations*. San Francisco, CA: Sierra Book Club.
Margulis, L. (1998). *Symbiotic planet: A new look at evolution*. New York, NY: Basic Books.
Marston, S. A., & Smith, N. (2001). States, scales and households: Limits to scalar thinking? A response to Brenner. *Progress in Human Geography, 25*(4), 615–619.
Massey, D. (1994). A place called home? In D. Massey (Ed.), *Space, place and gender* (pp. 157–173). Minneapolis, MN: University of Minnesota Press.
McLuhan, M., & Lapham, L. H. (1964). *Understanding media: The extensions of man*. New York, NY: McGraw-Hill Book.
Moran, E. F. (2000). *Human adaptability: An introduction to ecological anthropology*. Boulder, CO: Westview Press.
Poynter, J. (2006). *The human experiment: Two years and twenty minutes inside biosphere 2*. New York, NY: Thunder Mouth Press.
Preiser, W. (1991). Environmental design cybernetics: A relativistic conceptual framework for the design of space stations and settlements. In A. Harrison, Y. Clearwater, & C. McKay (Eds.), *From Antarctica to outer space* (pp. 147–160). New York, NY: Springer.

Ritsher, J., Kanas, N., Ihle, E., & Saylor, S. (2007). Psychological adaptation and salutogenesis in space: Lessons from a series of studies. *Acta Astronautica, 60*, 336–340.

Schutkowski, H. (2006). *Human ecology: Biocultural adaptations in human communities*. Berlin, Germany: Springer.

Steiner, D. F. R. (2002). *Human ecology: Following nature's lead*. Washington, DC: Island Press.

Stuster, J. (1996). *Bold endeavors: Lessons from polar and space exploration*. Annapolis, MD: Naval Institute Press.

Suedfeld, P. (1991). Groups in isolation and confinement: Environments and experiences. In A. A. Harrison, Y. A. Clearwater, & C. P. McKay (Eds.), *From Antarctica to outer space: Life in isolation and confinement* (pp. 135–146). New York, NY: Springer.

Suedfeld, P. (1997). Homo invictus: The indomitable species. *Canadian Psychology, 38*(3), 164–173.

Suedfeld, P. (2005). Invulnerability, coping, salutogenesis, integration: Four phases of space psychology. *Aviation, Space, and Environmental Medicine, 76*(Suppl. 1), B61–B66.

Suedfeld, P., & Steel, G. D. (2000). The environmental psychology of capsule habitats. *Annual Review of Psychology, 51*(1), 227–253.

Swyngedouw, E. (2004). Scaled geographies: Nature, place and the politics of scale. In E. Sheppard & R. B. McMaster (Eds.), *Scale and geographic inquiry* (pp. 129–153). Oxford, England: Wiley-Blackwell.

Tuan, Y. F. (1974). *Topophilia: A study of environmental perception, attitudes, and values*. New York, NY: Columbia University Press.

Tuan, Y. F. (1977). *Space and place: The perspective of experience*. Minneapolis, MN: University of Minnesota Press.

Vogler, A. (2004). *Change of paradigms: Designing habitats v. designing machines*. NASA Explorations Systems White paper.

Vogler, A., & Jorgensen, J. (2004). Windows to the world; doors to space: A reflection on the psychology and anthropology of space architecture. In *Space: Science, technology and the arts (7th Workshop on Space and the Arts)* (pp. 18–21). Noordwijk, The Netherlands: ESA/ESTEC.

Walton, J. R., & McConocha, D. M. (1996). Relational dimensions of dog ownership. *Psychological Reports, 79*(3), 947–950.

White, F. (1998). *The overview effect: Space exploration and human evolution*. Reston, VA: American Institute of Aeronautics and Astronautics.

Worringer, W. (1997). *Abstraction and empathy: A contribution to the psychology of style*. Chicago, IL: Elephant Paperbacks.

Chapter 4
Alternative Technological Interfaces with the Space Environment

4.1 Premise

To say that technology produces useful things understates the prodigious power that it has; technology is much more than the sum of all devices. It is not simply the making of gadgets or products of invention—it is a force majeure—with implications that are social, political, economical, cultural, and even evolutionary.

There is a wide range of views among scholars as to the benefits and detriments of technology. Positions at the extremes tend to ascribe too much agency to our inventions, and too little agency to its creators and users, or vice versa. But although technology has an impact that is greater than the individual artifacts themselves doesn't mean that we cannot control their influence or make thoughtful, deliberate choices as to how (or even whether) they should be deployed. A tempered and less deterministic position is more defensible, one that is cautious and inherently skeptical of extreme claims, but still optimistic in the *possibilities* for the future.

Moreover, the extremists' attempts to characterize technology as either innately good or innately destructive create a false dichotomy between humans and their machines, between the natural and the synthetic. Technology is not something external to human being but rather an extension of human being; it doesn't threaten humanity—on the contrary, it is integral to our humanity. Not only the things that we make, but also the capacity for making them, are just as much a part of the human condition as the mind and body. Our identity as well as our social relations are augmented through artificial means.

This is not to claim however, that the transformative power of technology always has a positive influence in our lives. The double-edged sword of technology can both empower us in some ways and debilitate us in others; instantiate new relationships while cleaving existing ones. Technology essentially offers us different opportunities. But having more options should not be equated with progress—the measure of progress depends on the goals we set for ourselves, and whether our technologies help us attain them.

E.S. Lockard, *Human Migration to Space*, Springer Theses,
DOI 10.1007/978-3-319-05930-3_4, © Springer International Publishing Switzerland 2014

In order to develop appropriate technological interfaces for living in Space, we must first determine what our objectives are. I have identified four specific areas of technology that will be significant with regards to the ability to adapt to the Space environment: cyborg technologies, artificial reality, terraforming, and robotics. Cyborg technologies are interventions that modify the body (subject) in some way; these include implants and prostheses, as well as genetic engineering. Artificial reality consists of technologies that create new non-material environments such as cyberspace and other virtual environments, whereas terraforming modifies existing ones. Robots and artilects constitute a new type of entity (other) that will have an increasing and indispensable presence in future missions to Space.

Each alternative in and of itself has its potential benefits and drawbacks; what makes the difference in large part is how we choose to implement them. Human intent, more than intrinsic technological capability, is the critical factor in the decision-making process, which is why I have formulated four alternative approaches based on objectives rather than outcomes. From a phenomenological perspective, how our technologies change the environment is less relevant to this project than how technologies change our perceptions of, and relationship to, the environment.

4.1.1 Literature Review

Technology is developing at an ever-increasing pace, outstripping society's ability to process and adapt to the magnitude of its impacts in every corner of human life. We have witnessed more sweeping changes through advances in technology during the last decade than we have in the half century that preceded it—and according to Moore's law, we can expect this rate to continue to accelerate (Kurzweil 2005). The speed with which new technologies are introduced places even greater stress on our ability to assimilate them into our social and cultural fabric. The realities never live up to the promise of new technologies, which are always touted in the most appealing terms when first introduced. This runaway momentum in high-tech development has subsequently divided many scholars on the subject, creating a polarized techno-partisanship. For some, rapidly advancing technologies cause anxiety, seeing them as a threat to our humanity.

The kinds of technologies that will become viable in the near future, and the immense transformative capacities they confer, is hardly under dispute, though there is considerable disagreement as to the range of applications those technologies should take. What seems to generate the greatest debate however, are the ontological ramifications of those technological developments. In the salvationist camp are the 'technophiles,' the most prominent of whom include Ray Kurzweil, Kevin Kelly, Hans Moravec, Asher Seidel, Nick Bostrom, and George Dyson. All of them subscribe to a fatalistic view of exponential technological growth that will elevate human life through its machines to new levels of complexity and capability. Kurzweil believes—based on Moore's law in which "the Singularity is the inexorable result of the law of accelerating returns" (Kurzweil 2005, p. 36)—that "by the time of the Singularity, there won't be a distinction between humans and technology. This is not because humans will have become what we think of as machines today, but rather machines will have progressed to be like humans and beyond" (Kurzweil 2005, p. 40). Moreover, in this view, humans are powerless to halt or even suppress this technological usurpation. According to Kelly, "the world of the made will soon be like the world of the born: autonomous, adaptable, and creative, but consequently out of our control" (Kelly 1994, p. 4). Eventually we could even achieve immortality through the transplantation of our consciousness into the hard drive of a computer which will then live on after our body dies (Moravec 1988). The technophiles embrace the literal image of the cyborg, the image in which humans and machines merge to become indistinguishable from one another, resulting in an entity that will transcend the individual aspects of the organic and the synthetic. The subtext of their advocacy is exultant in tone; there are no reservations or ambivalence about the direction of this inevitable technological radicalization.

The 'technophobes' on the other hand, such as Bill McKibben, Bill Joy, and Francis Fukuyama, tend to see the negative effects of new inventions, and fear that unfettered technological advancement will eventually precipitate the demise of humankind—if not biologically, then at least metaphysically. Arguing that our lives have become oversaturated with needless gadgets and devices, they eschew the GRIN (Genetic, Robotic, Information, and Nano-) technologies that they believe will take away our humanity. McKibben, whose arguments are reminiscent of luddite ideology, warns against the merger of humans and technologies that the technophiles so enthusiastically embrace: "Making devices of ourselves would be the logical end of our technological momentum; it would end the tension we feel between the real and the artificial. And that might be a relief; there'd be nothing nagging at us to go 'make more' of our lives. But that tension is the last remaining fully human part of us" (McKibben 2003, p. 5). Fukuyama echoes a similar sentiment, but his concerns revolve around what he sees as the political consequences of rampant technological development: "Human nature shapes and constrains the possible kinds of political regimes, so a technology powerful enough to reshape what we are will have possibly malign consequences for liberal democracy and the nature of politics itself" (Fukuyama 2002, p. 7). Both objections are derived from tacit assumptions about human nature and political dispositions that are highly contestable, however. Moreover, their arguments against high-tech are as much a slippery slope as those of the optimists, and beg the question: Which technologies should be deemed significant and which superfluous? If the opponents of expanding technology accuse its proponents of showing no restraint, then conversely, the same criticisms can be levied against them in their bias towards flatly dismissing all advanced technologies.

Though they occupy opposite corners of the polemical ring, what both sides have in common however, is an implicitly deterministic view of technology that diminishes human agency. From both perspectives, our machines hold sway over those who invent or use them, who simply react to them. Technologies are depicted as either intrinsically beneficial or intrinsically detrimental, regardless of how they are applied by its users. But debates over whether technology is inherently good or evil are overly reductive. It is more accurate to describe technology as transformative (Dator), a term which implies that human intentionality plays an indispensable role in assigning their value, whether it be positive or negative. “Determinism. . .does little justice to the genuine choices that arise, in both principle and practice, in the course of technical and social transformation” (Winner 1989, p. 10).

Occupying a more tempered but hopeful position between the two extremes are the ‘techno-agnostics’—Jaron Lanier, Andy Clark, David Brin, and Joel Garreau—who acknowledge the transformative power of technology, but are equally critical of claims of either immortality on the one side or annihilation on the other. From a skeptical perspective, agency is still retained by the human maker/user, and it is the maker/user who can ultimately determine the directions that technology can take us. In this non-deterministic view, technology can and should be controlled, but not suppressed. Thoughtful engagement, a negotiation of sorts, between humans and the things they produce underlies their agnosticism. Technological determinism sedates humanity into a state of dormancy and compliance. Technological ambivalence on the other hand, is actuating; it compels humans as thoughtful agents to both resist and envelop change. We confront and resolve our tensions in a process that is constantly stabilizing but never stable and dormant. It is this very activity, this ongoing arbitration with our technologies—and not the patent dismissal or unequivocal acceptance of them—that is central to being human. “Even if technology is advancing along an exponential curve, that doesn’t mean humans cannot creatively shape the impact on human nature and society in largely unpredictable ways” (Garreau 2005, pp. 205–206).. . .“technology does not control us, but we control technology” (Garreau 2005, p. 234).

4.2 Theories of Technology and the Relationship to Environment

4.2.1 Technology as Activity and Artifact

A pervasive conceptualization in the literature characterizes technology not just as products/inventions/devices, but as process/activity/experience/phenomena/force (Kelly, Kurzweil, McLuhan, Arthur, T. Hughes, Basalla, Winner, Haraway, Hillis, Mander, McCarthy and Wright). “Technologies are not merely aids to human activity, but also powerful forces acting to reshape that activity and its meaning” (Winner 1989, p. 6). Though inventions play a mediating role between humans and their environment, it is perhaps more accurate to describe technology in broader terms as a constitutive or generative force that is the catalyst for ontological transformation. Whereas posthumanist and cyborg literature posits technology as an extension of the human *subject*, McLuhan makes the claim that “any technology gradually creates a new human *environment*," characterizing environments not as "passive wrappings, but active processes" (McLuhan and Lapham 1964, p. vi). In other words, technology *instantiates* both new identities *and* new environments. It is not merely the information transferred by the technology that changes relations, but the very use of the technology itself. The arguments come to the same conclusion from different angles, the salient point being that there can be no definitive distinctions between technology and subject, or between technology and environment.

Our inventions—whether they be complex machines or simple hand tools—do not exist only as objective and independent artifacts but are embedded in, and shaped by, history and culture. The history of simple inventions like the wheel, for example, show that its emergence and its subsequent evolution were not as ubiquitous as commonly assumed. It was adopted in some regions for which it was functionally appropriate for the topography, and others not at all.[1] This indicates that technologies are not applied universally, but selectively depending on regional utility.

Hillis echoes McLuhan when he describes technology not as product, but as activity; in referring to David Rothenberg's circle of intent and result, he writes that the potential of a technology "seems realized, but its usage, both as a thing external to ourselves and as a mechanism for extending our reach, then suggests new intentions...[in turn] new technologies are imagined by scientists and the broader sphere of social relations...Rothenberg's circle of human agency and technical affect supports understanding of technology as an *activity*, as something that is done" (T. Hughes 2004; Hillis 1999, p. 33). Rothenberg's statement posits a recursive relationship between intention and invention that is reflected in Haraway's view. She writes that "machines can also be thought of as arrangements of social practices in which the stability of definition and relational positions between humans and machines of necessity must remain somewhat unstable and fluid...To speak of a world as congeries of practices doesn't mean that all the actors are human" (Hillis 1999, p. 34).

This commonly held view of technology suggests that there is the primary level of activity associated with the *intention* of an invention—the function it was designed to perform—as well as a secondary level of activity associated with the *effects*—generally unforeseen or unintended—on the population at large. It is at this second level where social relations can be transformed.[2] Hillis implicitly acknowledges these two levels of activity in his distinction between 'technologies' and 'tools': technology constitutes "a complex assemblage of many devices and therefore [is] not only a tool. Technology organizes labor practices, in contrast to a

[1] Basalla cites that in regions defined by rugged rocky landscapes, the wheel did not become a prominent mode of transport; a camel was far more capable of negotiating irregular terrains. The wheel was only practical in regions where topography was relatively smooth (Basalla 1989).

[2] McLuhan cites, for example, the advent of electricity as a decentralizing social force, one that in blurring distinctions between night and day, radically and irrevocably altered social interaction—though it was not conceived for this purpose.

person using a tool to extend his or her grasp or power. . .Technology, therefore, is not only gadgets, mechanisms, and tools, but increasingly, also a set of social practices depending on distributed knowledge and skills" (Hillis 1999, p. 34). The distinction of primary and secondary levels of activity speaks to the reflexive relationship between technology and human beings that McLuhan describes in *Understanding Media*; just as changing technologies transform society, so does a changing society transform technology. This dialectical nature thwarts theories of technological determinism on one extreme and social constructivism on the other.

Contrary to constructivist arguments, technology cannot be characterized as neutral or value-free. Our inventions have a history; the assumptions, preconceptions, and biases of its designers are embedded in it as well. Mander writes about the inherent character of technology and its impact on second-order (meta-)activity:

> . . .one hears the same homily: 'The problem is not with technology itself, but with how we use it, and who controls it.' This idea would be merely preposterous if it were not so widely accepted, and so dangerous. In believing this, however, we allow technology to develop without analyzing its actual basis. . .A prime example is nuclear energy, which cannot possibly move society in a democratic direction, but *will* move society in an autocratic direction. Because it is so expensive and so dangerous, nuclear energy must be under the direct control of centralized. . .institutions. Community control is anathema. . .it is fair to say that nuclear technology *inherently* steers society toward greater political and financial centralization, and greater militarization.
>
> Solar energy, on the other hand, is intrinsically biased *toward* democratic use. It is buildable and operable by small groups, even by families. It does not require centralized control. It is most cost-effective at a small scale of operation, a reason why big power companies oppose it. And solar power requires no thousand-year commitment from society. So where nuclear energy requires centralized control, solar energy functions best in a decentralized form. These attributes are inherent to the technologies and reflect the ideological basis of each (Mander 1991, pp. 35–36).

However, a predisposition towards a particular use or social practice doesn't preclude an element of human control or intentionality in terms of whether, or to what degree, a technology should be applied. When deciding what kind of technologies to employ for the purposes of Space habitation, we must consider not only the operational/instrumental features of a particular technology (the engineer's perspective), but also how they affect human interaction, human perception, and human identity (the social scientist's perspective). For instance, how will regolith shielding influence the way in which inhabitants of the Moon relate to the lunar environment? How might improved communication capabilities with family members back on Earth affect relations between mission astronauts? How does isolation from the Martian environment in order to prevent cross-contamination impact crew members' sense of belonging?

When we fail to acknowledge the immanent characteristics of a technology, we abdicate our responsibility to guide (or limit, as the case may be) their development, and hence relinquish some degree of our agency. If we subscribe to a deterministic view of technology—that technology either unconditionally improves our lives and leads to progress, or conversely will destroy humanity—then we become 'technosomnambulists' (Winner 1986) by surrendering ourselves to the fate our technologies impose upon us. However, technological determinism does not follow from inherent predispositions, nor does the reflexive character imply that the manner in which technologies are employed is entirely attributable to the end user. The reality cannot be reduced to either extreme; a more comprehensive view of the role of technology slides along a continuum between diametric views.

4.2.2 Technologies of Expedience vs. Experience

Though specific technologies cannot be simply labeled as 'good' or 'bad', the inherent traits of a technology—as Mander argues—predispose them to certain uses which can influence social and political relations. There is a subset of devices that are expressly conceived to make our lives more convenient, more streamlined. Implicit in this category of inventions is the supposition that the resistant elements of our environments, both physical and social, are obstacles to be minimized or bypassed altogether. These types of technologies that do not confront or engage social or material resistances can be characterized as technologies of expedience. Though designed with the intent to increase efficiency (primary activity), they also effectively inhibit exchanges with our environment (meta-activity).

Inventions that allow us to interact with the physical world in a way that would otherwise be impossible without them are defined as technologies of experience. These inventions are qualitatively different from those of expedience in that they give us opportunities to establish relationships with our natural world that did not previously exist. Early technological artifacts engaged us with our natural surroundings, allowing us to develop more proactive and participatory relationships with our environment. In being less at the mercy of the capriciousness of Nature, we could reap greater bounties from our environments. Because of the more direct and immediate relationships with the natural world that those early technologies cultivated—because it *enabled* labor through engagement with the land rather than circumventing it—there was a deeper connection to, and sense of gratitude for, what the physical world could provide to sustain life.

Technologies that instantiated layered relationships made possible what was previously impossible, and thus engage its users with their world on a level that could not have existed without them. An example of a technology of instantiation is the first boats that were built which could take us across the seas to an undiscovered land. Navigated by the stars and powered by its occupants, these ships allowed their passengers to be fully immersed in both the peril and the wonder of traveling across water. In contrast, technologies of expedience merely facilitate what was already possible, and focus only on improving efficiency. (Hydrofoils or cruiseliners are examples of technologies of expedience which effectively remove us from the experience of the sea.) This mediation, which emphasizes convenience over confrontation, creates phenomenological distance between us and our external world rather than integrating us with it.

Contrary to the position of both Mander and McLuhan, the inherent nature of a technology is not always fixed or definitive. Dimendberg cites the development of the automobile (and subsequently the highway) as an example of technology that was intended to possess both a functional as well as an aesthetic dimension. As a technology of performance it served as a mode of transport, but as one of aesthetics it provided exhilarating sensory experiences of both speed and the landscape. (The Autobahn was in fact, designed to re-present the bucolic German countryside to the driver, framing views and vistas specifically with the traveling spectator in mind.) As a mode of re-presentation, the automobile allowed us to see our world anew and on a much larger scale, making for a richer, more tangible experience. The automobile captured stillness in motion (Dimendberg 1995). As McLuhan observed, eventually the auto became mundane and the aspect of excitement disappeared; it became simply a means by which to expedite our journey to some remote destination (McLuhan and Lapham 1964). Travel ceased to be an ends in itself, an experience to be savored. No longer a means of *re-presenting* the landscape, the disengaged experience of the countryside reduced it to a mere *representation* of the landscape, and as a result the experience of travel by car changed to one of active alienation from the physical world. Time spent in transport is no longer lived time, nor is the space between destinations lived space. The automobile changed from a technology of instantiation to a technology of expedience. Highways are the ultimate example of nether-space, a veritable spatial purgatory, serving merely as a connector between places, but never a place in itself. Thus the time spent on highways is spent nowhere; during transport we are both placeless and timeless.

Technology reached a turning point with the Industrial Era. The revolution in industrial technology gave rise to the modernist movement in which the machine was exalted. Modernism was the eloquent expression of the Newtonian worldview—rational, mechanistic, and efficient. Architecture, in the words of Corbusier, became a 'machine for living.' Humans no longer had to work directly with Nature, but could circumvent it, allowing the *machines* to work more efficiently with Nature instead. Automation 'emancipated' humans from their involvement with the natural world by introducing another kind of mediation. The assembly lines of mass production meant that humans had no direct connection with the fruits of their labor. Through these changes a new cultural ethos emerged, one imposing a standardized, mechanistic efficiency over the idiosyncrasies of individuals. Our social relations, as well as our relations to our environments were radically transformed by this technological revolution.

In this new era of modernity, there was a shift from enriching technologies of experience—which *established* new and different relationships with our environment—towards reductive technologies of expedience, which were intended to *bypass* relationships for the sake of increased efficiency. In referring to the lust for efficiencies, Hegel believed that technological devices strip things of their 'soulful interiority,' resulting in an "ever reproducible exteriority" (Danius 2002, p. 195). The expedience afforded to us by our modern technologies also fundamentally altered the way we perceive the natural world. Whereas technologies of experience engaged and activated its users, technologies based on efficiencies have led to devolved modes of perception, disengaging and alienating us from the resistant material world. In our insatiable desire for speed and increasing mobility, our technologies treat time and space as obstacles to be overcome, not as something to be relished. We live not 'in the moment' so to speak, but in the instant, severed from a continuity to the past and the futures. The result is the collapse of space and time. Lived experience is no longer something to revel in but to streamline. Efficiency strives to relieve us of our burdens through the elimination of resistances; experience is rooted in meaningful negotiations with the physical world. In the quest for efficiency over experience, for conquest and control over confrontation, we have succeeded in dominating our environment, but in the process we have alienated ourselves from the world around us.

It will be tempting, and sometimes imperative, to employ technologies of expedience in Space. Many of the conditions that humans will be exposed to once they leave the confines of Earth are life-threatening, and so it is logical to either minimize our exposure to them or avoid them altogether. If we look at possible interventions only from a life-safety perspective, then this direction would seem to make the most sense. But I would argue that an expedient strategy presents another set of dangers when understood from the perspective of adaptation. Encountering and negotiating environmental resistances—which comes only through the experience of our surroundings—is essential to the phenomenological aspects of the adaptation process. Technologies that are used to protect us from the dangers of our environments will also need to be supplemented with technologies that are geared towards promoting and enhancing experience of the physical environment as a means of developing familiarity, rather than towards expedience, which will only generate or exacerbate feelings of alienation towards the environment.

4.2.3 Technologies of Artificial Reality

> In the society of spectacle, the idea has become the image and the real is the imaginary.—Julian Stallabrass

Stallabrass' statement makes clear the ontological conundrum that cyberspace presents, as it contrasts to physical space. Cyberspace is often described in perplexing phrases such as "space without space, a 'non-place'" or "a space of flows characterized by timeless time and placeless space" (Dodge and Kitchin 2001, p. 62). Artificial realities create a parallel universe through our imagination—a world that is devoid of the body itself—in which representations are the only things that exist. But humans cannot be reduced to digital simulations and mental processes; embodiment is essential to our sense of identity, to our relations with other people, and to our relationship with the physical environment. "The real and embodied places of the world—and their synthesis of meaning, nature, and social relations—are central to grounding self-identity and organizing sociocultural practices in a coherent fashion...VEs seem to subvert this resistant materiality...They suggest that the lived world need not be embraced but simply reprogrammed until it matches 'our' desires" (Heim 1998, p. 203).

Virtual environments and cybertechnologies challenge the ontology of identity and social relations like no other technologies that have preceded it. Many theorists of cybertechnology (Stone, Dibbell, Boellstorff, Ondrejka, and Lanier) argue—in the Cartesian tradition where the mind is primary to existence—that despite the absence of physical presence, virtual reality defines 'real' space, our avatars or online personae (as extensions of our subjectivity) are 'real' people, and the social interactions that occur within these environments constitute 'real' community. They believe that "computers are arenas for social experience and dramatic interaction...inside the little box are *other people*," (Stone 1996, p. 16) reflecting a view of subjective experience that privileges mental perception over physical sensation. Lanier even goes so far to suggest that "virtual house building will be a part of a (future) communication without codes. 'If you make a house in virtual reality...you have not created a symbol of house or a code for house. You've actually made a house'...[but Hillis mocks this sentiment by arguing that] Lanier's proposals will do little to alleviate housing shortages, substandard accommodations, or homelessness...Lanier intends his vision to be liberatory. He fails however, to address the issue that VR trades in images of real things, as well as in simulations with no referents" (Hillis 1999, pp. 191–193). It may very well be perceived as a house, and even possess phenomenological aspects of home, but the cyber-instantiation denies its material aspects altogether.

Though the advent of technologies which provide virtual experience through artificially constructed realities have allowed us to occupy non-present temporalities as well as non-physical spatialities, they also remove us from real space and real time, hence creating a condition of disengagement with the natural world. Current virtual technologies actively suppress our physical dimension, which can only result in a state of alienation. While cybertechnologies present valid alternative conceptions of space, identity, and environment that Hillis may have been too quick to reject, they have also neglected the resistances and constraints of the physical world that have shaped our conceptions in the first place. "Spatialization of [our] consciousness...minimizes the environmental evidence of the places of the Earth...it is through recognition of [relational] continuities [of phenomena grounded in real places]—exactly what the promoters of VEs suggest that users cast aside—that we achieve external confirmation of our individual and group identities" (Hillis 1999, p. 189). VRs offer us another level of interaction, transforming identity, space, and social relations; but also dispenses with those aspects which are derived from physical experience. Hayles asserts that people "have something to lose if they are regarded solely as informational patterns" (Hayles 1999, pp. 15–16). Virtual technologies can *augment* but cannot *supplant* identity and social relations grounded in physical space. As Heim puts it: "VR transubstantiates but does not imitate life" (Heim 1998, pp. 48–49). Subjectivity must have physical presence, and through it, sensuous experience of its environment.

VRs represent a kind of ontological purgatory which is problematic on many levels. No single position in the discourse captures the nature of VR in totality, and likewise, none of them can be totally discounted. This state of interstitiality is reflected in the writings of Dodge and Kitchin: "Virtual landscapes are liminal spaces. . .neither here nor there, neither imaginary nor real, animate but not living or dead, a subjunctive realm wherein events happen in effect, but not actually" (Dodge and Kitchin 2001, p. 62). The absence of constraints that establish limits or resistances for our negotiations in the physical world presents unique problems for the development of social relations in cyberspace; "the extreme mutability and multiplication of identity possible in cyberspace collides with the desire to build communities based upon honest communication. . .it might form the ultimate solipsism, because space and perception are infinitely malleable to the user" (Stallabrass 1995, p. 16).

Another danger is that VEs, in relegating the body to a subordinate status with regards to identity—or even as an obsolete artifact—decouples subjectivity from the body. The secondary level of activity of cybertechnologies can thus be characterized as both a disembodying and despatializing force. In cyberspace, identity is reified exclusively through communicative exchanges. There is no encounter with a physical body; presence is asserted by discontinuous, fragmented, and often capricious informational representation. Materiality and its concomitant resistances are not essential to social relations in cyberspace, and thus the body becomes dispensable in that process. Despatialization, or the *negation* of physical space, is very different from the collapsing of space attributed to technologies of speed in which physical space still exists but is *compressed* or *dislocated*. Cybertechnologies, in reducing space to metaphor, and occupied (so to speak) by non-material constructions of identity, actually *annihilate* physical space. If there is no space there can be no body, and without body there is no space. Hillis argues that "this relocation of what is concrete to an imaginary or metaphoric space assumes that the act of communication is a wholly adequate substitute for embodied experiential reality; it exchanges communications technologies for the reality of places and dispenses with, for example, empiricism's concerns about sense data and how things are understood as true and/or real. . .finite physical bodies are thought secondary to our minds and representational forms—a dynamic that is built into virtual technologies" (Hillis 1999, pp. xv–xvi). Biocca adds that "instead of a mind communicating through a body to another body, we have two disembodied conversations, a sterile coupling of abstract symbol generators" (Biocca 1997).

Heim also acknowledges the distinction between technologies of expedience and experience when he argues that we are now at a point where "virtual world design faces the crossroads: the tunnel or the spiral. The tunnel sucks us further into technology as a forward-thrusting obsessive fixation. The spiral. . .returns us to ourselves, repeatedly deepening the awareness we enjoy as primary bodies. Technology can string us out to become peripherals of machines, or it can add another layer to the ever-deepening return to enriched, enhanced existence" (Heim 1998,

p. 76). Despite its prophetic overtones, Heim is claiming here that we must make this decision to either allow technology to subjugate us or use it to empower us—technology does not (or should not be allowed to) make this decision for us.

While there are many necessities and benefits to employing virtual technologies in Space, we must use them judiciously. The problems associated with cyberspace are only compounded by its parallels with outer Space. Like cyberspace, which is "free of the context of geographical place and community" (Dodge and Kitchin 2001, p. 53), outer Space has no terra firma, no grounding or orientation, no reference. Cyberspace is described metaphorically as a dark void; outer Space is de facto a dark void. Physical bodies do not occupy cyberspace, but though the body still exists in outer Space, its presence is greatly reduced by the minimizing effects of micro-gravity. Without location of body, cyberspace cannot convey a sense of place; in outer Space "there's no there there" (Dodge and Kitchin 2001, p. 62). Distance is meaningless in cyberspace: it is represented either by zero (here) or infinite distance (nowhere). Outer space likewise represents the infinite.

4.3 Levels of Technological Intervention Based on Exploration Objectives

Since intensive technologies will be an integral aspect of life in Space—even more so than on Earth—the nature of technological mediation will have a profound influence on how we perceive our new surroundings. The habitats we construct can be experienced either as functional and practical shelter, or as a dwelling where we feel a sense of belonging and connection. Our approach to designing them will be informed by our attitudes towards the Space environment.

Technologies deployed for life in Space can be classified according to the different objectives listed in Table 3.1 (survival, performance, habitation, or adaptation), with each level building on the requisites of the previous one. The first and second levels of technologies has been in use since the early years of Space exploration, while the third level—which addresses the limited scope of habitation—is currently in development. However, the fourth level, which deals with evolutional adaptation, has yet to be considered. Evolutional factors are not yet even part of the discourse on habitability.

4.3.1 Level I: Survival

Habitat technologies that insure survival—which include life safety systems, airlocks, CELSS, pressurization, radiation shielding, oxygen and water recycling, heat insulation, etc.—are of an immediate, constant, and necessary nature. Because the singular focus is on human preservation, they are geared to maximizing safety and are non-negotiable; therefore, little or no demands are made on the part of the individual.

This has been the primary priority and the only human factors criterion for the first three decades of manned Spaceflights. Comfort was not a consideration, given the short stints the crews spent in Space. Little activity was required on the part of the astronauts, so performance demands were at a minimum. But as durations in Space grew longer, and as other functions became part of the mission, additional technologies were incorporated, though those still fell within an engineering scope.

4.3.2 Level II: Performance

As scientific research became a larger part of the agenda of spaceflight, technologies that were conducive to work performance received more attention. In optimizing conditions to conduct scientific experiments, to repair and maintain machinery, and for general operations, the emphasis was on functionality, efficiency, and practicality. Technologies were in service to the human operator, but only insofar as they assisted in getting a task done. Work activities and work-related objectives took precedence over welfare and comfort. Tools, pressurized spacesuits, helmets and gloves, velcro fasteners, tethers, and samples canisters, fall into the category of performance technologies. Their sole purpose is to accomplish a particular task, and has nothing to do with immediate survival, habitability, or adaptation to the Space environs.

4.3.3 Level III: Habitation

Technologies that are applied towards habitability fall into two sub-categories:

1. Technologies required for physical wellness, which include artificial gravity, simulated daylighting to regulate sleep cycles, and physio-training equipment to help reduce muscular atrophy and bone loss. Because these conditions are not life-threatening, they tend to be less urgent than Level I considerations, and can be subject to partial acclimatization on the part of the individual. Devices need only be employed intermittently and to varying degrees, to be effective. (For example, centrifugal mechanisms to simulate gravitational forces can be turned on or off, or be calibrated to induce partial-gravity only.)
2. Technologies that address psychological and social well-being such as acoustics, privacy, communications, recreation, and mitigation of isolation and confinement. They too are also less immediate and need not be constant, but are equally indispensable for long durations.

Both sub-categories focus on the well-being and quality of life of the individual and the crew as a whole, with particular emphasis on human comfort, but issues of longevity and durability are not included in this scope.

4.3.4 Level IV: Adaptation

Technologies that promote evolutional adaptation are those which engage us with our physical surroundings rather than only shelter us from them, and which help us to develop aesthetic or meaningful sensibilities of place (phenomenological), as well as corporeal affinities (physiological). Because this level fosters engagement, it excludes technologies that bypass experience and are expressly intended for efficiency, expedience, or convenience. Examples of this category are remote sensing devices that can be used to provide an enhanced sensory experience of the environment which may not otherwise be possible, or pressurized spacesuits used for EVAs.

The nature of this type of technological mediation however, may directly conflict with those of short-term survival and long-term comfort in that they are designed for the purpose of negotiation with the resistances of the natural environment. Radiation shielding on both the Moon and on Mars will be necessary to protect us against cosmic rays and solar flares—but they will also inhibit the experience of our surroundings. Ultimately we will need to use technologies not only for the purposes of survival and comfort but to better integrate us into our new Space environment. Technologies that promote robustness, resilience, and stability, will be essential for human longevity beyond Earth.

Whereas the timeframe for survival, performance, or habitability extends to the lifetime of an individual, the timeline for adaptation exceeds the lifespan of an individual, and therefore the focus is not exclusively on the welfare of an individual for the duration of their life, but also to insure the prosperity of the entire human race over generations.

The next step is to examine how the range of current and emerging technologies could be implemented to support long-term adaptation to the Space environment, including possible specific strategies.

4.4 Proposed Technologies for Adaptation in Space

The purpose of this section is not to inventory all the possible technologies that will be employed in Space but to focus specifically on those technologies that could serve to promote human adaptation to Space. Intentional human-technology relations can be described as falling into three categories: embodiment relations, hermeneutic relations, and alterity relations (Coeckelbergh 2011). Embodiment relations refer to the amplification of bodily perception in which technology comes to be experienced as an extension of ourselves, and not as something distinct from it; cyborg technologies, which include prosthetic or implanted devices, fall into this category. Hermeneutic relations refer to the role technology plays as a mediator between subject and physical world that shapes interpretations of it; remote sensing and teleoperation are examples of virtual technologies that change the way we understand and experience our environments. And finally, alterity relations refer to the technological object perceived as an other, a relation in which the technology possesses a degree of autonomous subjectivity; robots and androids constitute these types of relations.

4.4.1 Cyborgism

The term ‘cyborg’ was first coined in 1960 by scientists Manfred Clynes and Nathan Kline to describe a human who, through the incorporation of “exogenous components extending the self-regulatory control function of the organism” (Clynes and Kline 1960, p. 31), could be adapted to living in Space. Rather than creating artificial environments suitable for human habitation, why not create artificial humans suitable for Space environments? The pair of scientists reasoned that “if [humans] attempt partial adaptation to space conditions, instead of insisting on carrying [their] whole environment along with [them], a number of new possibilities appear” (Clynes and Kline 1960, p. 30). The methods they proposed in order to achieve this relied primarily on pharmaceuticals, some of which were already available and others speculative. At the time of their far-fetched proposal, genetic engineering was not yet possible, or even foreseeable. But today, as technology comes closer to making that idea a feasible one, it still hardly garners much consideration. Despite that it would make human Space travel a much less risky venture, the radical alteration of human bodies is a much more disturbing proposition that alteration of the environment.

Though the concept of the cyborg is only 50 years old, it is a process that actually began long before we dreamt of traveling to Space. As exogenous devices have been implemented to extend our capabilities since the earliest days of our existence, we have always been cyborgs in degree; even the earliest and simplest technologies were expansions and supplementations of ourselves. The only differences are that in the present time there are many more options at our disposal, they have been invented at a much faster rate, and the extent of modification they promise is much greater. As technologies are implanted, embedded, attached, connected, and fastened to the body, we have become more aware of the complex symbioses between our organic selves and our inorganic devices.

“High-tech” examples that we typically associate with cyborgism include among many others: the Dobelle eye that can transmit images directly to the brain via camera, effectively bypassing the sense organ of the eye itself; Stelarc’s exoskeleton made of six additional mechanical limbs harnessed to the original two-legged body; and Bach-y-Rita’s TVSS—a prosthetic device controlled by the tongue

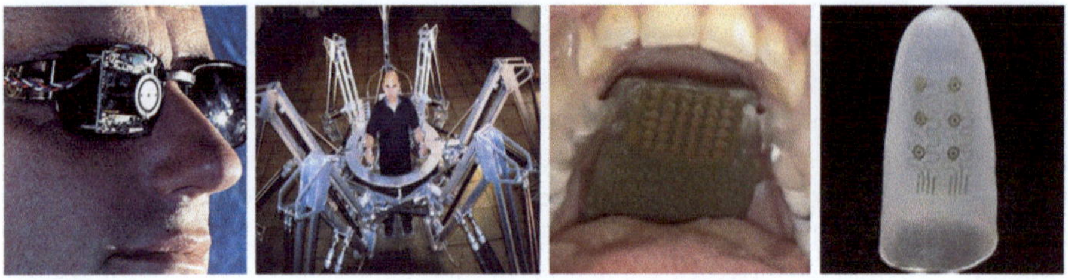

Fig. 4.1 "Cyborg" technologies: Dobelle eye; Stelarc's exoskeleton; Bach-y-Rita's TVSS; surgical fingertip

through tiny sensors (Fig. 4.1). But cyborgism is also defined by "low-tech" examples that we typically associate with disability rather than augmentation: the prosthetic leg which replaces the missing limb; the blind person's cane; the wheelchair which allows for mobility that the body can no longer provide for itself.

The machine called VGo (Fig. 4.2) is often thought of as a robot because of its apparent autonomous motility, but it is actually more a hybrid of robot, cyborg, and virtual reality. VGo is a mobile device with a camera, microphone, and speakers that allows a child to 'attend' school, though he is not capable to do so on his own. It can be described as a robot because of its independent freestanding form and capacity for movement; it can also be described as part of a cyborg system in that it extends the senses and 'presence' of the student who operates it; and it can also be described as a virtual reality in that the student experiences a remote environment that he does not physically occupy.

However, definitions notwithstanding, VGo is experienced more as a cyborg apparatus in that the other classmates see the device as the *student himself*, not as a *representation of the student*. Because VGo is approximately the same height as the students and the sensory organs are at the same level as its human counterparts, the classmates can speak and interact with him as if he were physically there. Children, who have more exposure to computers than their parents did, are not in the least uncomfortable with the ambiguity between the mashed organic and the mechanical entity.

Once the rest of society has come to embrace the increasing intimacy we have with our technologies, there are several emerging directions that we can, and should, pursue as we advance towards habitation in Space; these fall into several categories:

- Neurochemical (i.e., pharmaceuticals)
- Prosthetic devices which are extended from the surface of the body (e.g., limbs, sensory devices, mechanical apparatus)
- Body implants which are embedded within or below the surface of the body (e.g., neural implants, sensory organs, pacemakers, organ transplants, computerized micro-sensors, etc.)
- Remote devices (telepresence, teleoperation, remote sensing)

Fig. 4.2 Second grader attends school with his classmates

- Genetic engineering (i.e., somatic therapies which affect only the individual, or germline therapies which are hereditary)

Cyborg technologies could be geared to address specific environmental stressors of Space, altering the human physiology in order to:

- Develop resistance to radiation and solar flares so that cellular structure and function are not excessively disrupted
- Adapt to reduced gravity conditions to minimize or eliminate problems associated with bone decalcification, muscle atrophy, and fluid re-distribution
- Adjust vestibular/proprioceptive orientation in micro-gravity
- Alter the body's 24-hour circadian rhythms so they can align with either shorter or longer diurnal (light/dark) cycles, as well as allow longer periods of wakefulness during intense periods of activity and longer periods of sleep for trips to distant destinations
- Withstand much colder temperatures so that the ambient climates of Mars would fall within our somatic adjustment limits
- Eliminate the need for breathable air (where either none or a different composition exists) through oxygenation of blood and elimination of carbon dioxide from the body that does not require the use of the lungs
- Tolerate reduced atmospheric pressure levels at which internal fluids would normally vaporize
- Amplify existing sensory capabilities: improve eyesight to see further, finer, darker, lighter, or even beyond the visible light spectrum; smell to detect airborne toxins; taste to be more sensitive to a more limited, or different range of favors; tactile sensitivity to better distinguish materials by touch; hearing to allow for greater range of frequencies, lower decibels

Because the conditions of any extraterrestrial environment lie far outside our 'field of operation' (Clynes and Kline 1960), the technologies associated with migrating to Space will be nothing less than intensive, and many decisions we face will fall outside society's comfort zone—cyborg technologies perhaps most of all, because they will modify who we are so pervasively that it will cause us to question what it is that makes us human. Accompanying these issues are difficult ethical questions which need to be addressed: What types of technologies should be implemented, and to what extent? Should we favor temporary or removable prosthetic extensions because they are more palatable than permanent or invisible implants? Even if they do not alter our genetic makeup, will these changes make us less human—or even non-human? And if so, is that necessarily undesirable or unavoidable? Genetic engineering is already being applied towards developing organisms that could thrive in the Martian environment. Once these have been successfully tested, perhaps this precedent will help facilitate the transition to human acceptance.

4.4.2 Virtual Technology and Virtual Environments

Because of the dangers of direct physical exposure to the Space environment, virtual technologies such as remote sensing and teleoperation through robotic devices can furnish viable operational interfaces between the human crew members and the planetary environment. Several generations of MERs (Mars Exploration Rover) have been employed for scientific research (e.g., reconnaissance of the planetary topography and geotechnical analyses) through the use of semi-autonomous robotic rovers, allowing humans to experience the Martian landscape vicariously. The rovers effectively act as extensions of their operators' visual, tactile, motile, and cognitive abilities.

All technologies that either offer a remote experience of the physical environment through extending an individual's sensory, physical, or intellectual capacities, or otherwise create non-physical, digital environments, are in the broadest sense virtual technologies. VTs are nothing new; they have been around already for a long time. The difference however, with the newer generation of VTs is their degree of interactivity and engagement. The advent of the telephone allowed two-way voice communications in real time in which the callers' 'presence' could be experienced in two remote areas. Recent Skype software intensified the sense of presence by adding a visual component, in which faces, bodies and the spaces they inhabit can be seen. The visual dimension brings us a little closer to the *experience* of actually being there. In the future it is foreseeable that we will achieve full telepresence through an interactive medium that engages all the senses—not just sight and sound, but also smell, taste, and touch—as well as involves movement for the body to experience material resistance from afar. As VTs continue to develop towards providing fuller, richer experiences of the physical world, the distinction between the virtual and the real will become increasingly blurred.

Virtual environments are a type of immersive technology that can either simulate real, physical environments (the familiar) or create entirely new, imagined ones with no reference to physical reality whatsoever (the exotic). Because they are non-physical environments, users can escape the limitations of the biological body and the constraints imposed upon it by the resistant, material world. Their appeal resides in replacing the mundane and familiar with the exotic. Ironically, VEs in Space habitats might be more valuable in replacing the exotic with the mundane and familiar. Like many immigrants, the first Space settlers will no doubt feel a longing for their former home, and so as a means to offset feelings of isolation and confinement, as well as the understimulation that often accompanies these conditions, it will be tempting to use VEs to recreate natural landscapes of Earth: pastoral countrysides with chirping birds flying overhead, the ocean pounding along shorelines of fine sand, or the rustling of leaves as a breeze blows through a dense, lush rainforest (Fig. 4.3).

However, if VEs only offer comforting, nostalgic experiences of the familiar geographies of Earth, they can become a means by which to avoid the experience of Space landscape or escape from the physical realities of their situation. This will only lead to feelings of alienation and disengagement, and even perpetuate an implicit hostility towards their surroundings. By inhibiting the development of familiarity that occurs through degrees and durations of exposure to the planetary environment, virtual 'trips' to Earth will serve as a constant dismal reminder that their new Space abode is not their home but rather a suspension of dwelling and a state of disenfranchisement. Virtual technologies nevertheless, will be pivotal to the broader adaptation process, but its effectiveness will be highly contingent upon how, and for what purpose, they are implemented. Nostalgic uses must be supplemented (and eventually replaced) with exploration of the Martian landscape through devices like remote television goggles or self-controlled rover cameras which will promote stronger empathetic connections through facilitating greater familiarity with the local environment.

When we think of 'being in Nature' on Earth, we think of experiencing a place that is natural, and not mediated by gadgets, machines, or other devices. However, 'being in Nature' on Mars or on the Moon cannot be experienced without various media. VTs will necessarily frame and shape our interpretations of these encounters. Heim observes the irony of this relationship: "Cyberspace offers the allure of nature on another and different level. In a strange way, cyberspace restores a surrogate nature, reinstating the affective life at a level where the psychic framework of cyberspace absorbs nature. Cyberspace absorbs nature not only by transposing space exploration into electronics though telepresence, but by actually reviving certain experiential qualities that the human being once felt in nature and now feels in cyberspace" (Heim 1998, pp. 152–153)..."Nature, it seems, including the furthest reaches of the galaxy, is being captured and bottled in information space" (Heim 1998, p. 149).

As Space agencies begin to address the wider scope of human factors, VTs may also be designed—beyond robotic reconnaissance and information-gathering—to address habitability issues (e.g., Earth simulations in virtual environments or

Fig. 4.3 Virtual panoramic "window" on board the space vessel *Prometheus*

telecommunications with distant loved ones). While this could be beneficial to the overall short-term well-being of the crew by alleviating psychological and social stresses—which would in turn better insure success of long-duration habitation—it could also eventually become a mechanism of maladaptation in the long-term if empathetic connections to the Space environment are not cultivated. The acquisition of information about the climatic and geophysical properties of a planet is not sufficient alone to foster an affinity for it. Interface technologies must have an experiential dimension to them. Lanier warns that "information is alienated experience. . .experience is the only process that can de-alienate information" (Lanier 2010, pp. 28–29). If the primary purpose of the built habitat is to keep its occupants safe by buffering them from the 'hostile wilderness' of Space, it becomes even more critical that VTs establish an experiential connection to the extraterrestrial landscape through enhancing sensory relationships with it. VEs can instantiate a conduit not only to *learn* about the quantitative properties of the environment, but also to *experience* its qualitative properties in ways the body cannot experience directly. Rather than thinking of VTs as circumventing the body's capacities, we should think of them as extending or augmenting them.

There are currently two types of immersions available: the head-mounted display (HMD) and the cave automatic virtual environment (CAVE). The HMD creates a virtual world in front of your field of vision. With the exception of movements of the head and eyes, the body is not actively engaged; in fact "immersion results from the primary body's blackout for the sake of the cyberbody, and a tunnel-like perception of the virtual world is the consequence" (Heim 1998, p. 101). Although it is characterized as a virtual environment, it is closer to what Biocca refers to as a non-distal 'imaginal' environment (one that is mentally constructed, such as hallucinations and daydreaming) (Biocca 1997). The reconfirmation of embodiment through VTs is essential in orienting the subject in space; "a disembodied subjectivity messes with whereness. In cyberspace you are everywhere and somewhere and nowhere, but almost never here in the positivist sense" (Stone 1996, p. 398). One cannot apprehend place without the body. Dodge and Kitchin point out that "cyberspace becomes placeless. . .there is no physical place in cyberspace, only individual digital traces" (Dodge and Kitchin 2001, p. 63). And without a sense of place, "a 'psychic homelessness' informs VEs" (Hillis 1999, p. 207).

The CAVE, on the other hand, creates a real physical space (albeit finite) within which one can move about and encounter other individuals. This is what Biocca would call a proper 'virtual' environment because it is distal mediated. (The third type of environment is the physical one, which is distal immediate.) Unlike HMD, immersion actually depends on the use of the primary body. CAVE has what Heim describes as an apperceptive quality; that is, it entails proprioception and self-awareness, and contextualizes sensory perception. "From the viewpoint of user phenomenology, the difference is one of the experience of the self. [HMD] supports a focused self and [CAVE] supports a wide-angle awareness." (Heim 1998, p. 104). HMD is entrenched in solipsistic experience, and therefore would not be recommended for isolated and confined living environments such as Space habitats. The CAVE on the other hand, would be more suitable as it supports relational experience and could help to alleviate feelings of confinement.

CAVE gets us closer to the concept of progressive embodiment, which is defined as "the steadily advancing immersion of sensorimotor channels to computer interfaces through a tighter and more pervasive coupling of the body to interface sensors and displays" (Biocca 1997). When our techno-ecological relationships in Space have come to the point of embodied, sensual integration, our experiences will no longer be characterized as virtual, but instead as merged, embodied, or fused. As they develop towards more enhanced and comprehensive sensory experience, VEs become in fact less virtual, shifting from simulation to re-presentation. Re-presentations of the landscape by the habitat architecture, by showing them anew, render it more visible (Harries 1998). If conceived as technologies of instantiation, VTs can enrich our relationships with the natural world instead of denying them; they can reveal to us aspects of our physicality that could not otherwise be experienced.

Virtual technologies have until now sought only to *resemble* real or corporeal experience, and it is the intentionality behind the application of technology which has caused alienation from both our essential physicality as embodied beings, as well as from the physical world in which we are situated. This is not however, inherent to the technology itself, but dependent upon the objectives for which it is implemented. VTs are not by their nature antithetical to the experience of the natural world; by developing more profound and comprehensive sensory relationships with the environment, they have the capacity to bring us even closer to Nature, to facilitate a deeper connection to the physical Universe.

4.4.3 Terraforming/Ecological Modification/Planetary Engineering

Terraforming refers to modification of the planetary ecology so that it can support human life. While the process would vary depending on the conditions of the planet, Mars is considered a potential candidate for terraforming because it lies on the edge of the habitable zone and has features which make it technically possible to initiate it.

There are three stages that would be required to make Mars a habitable planet. First, surface temperatures would need to be elevated. This could be achieved by covering the surface with a blanket of dark lichen which would reduce the planet's albedo (reflectivity), so that light and heat would be absorbed rather than reflected away from the surface. Other options to trap heat would be to introduce CFCs (chlorofluorocarbons) to create a greenhouse atmosphere, or deploy orbital mirrors to concentrate the sunlight on the planet surface.

Second, triggered by warming of the planet, the abundance of existing CO_2 (a greenhouse gas) in the very thin atmosphere (0.1% of Earth's) would begin to hold the heat, which in turn would release more CO_2 through sublimation (Gjerde 2000). As the density of greenhouse gases increases, UV rays from the Sun would be captured, creating a runaway greenhouse effect. The introduction of an atmosphere, coupled with warmer temperatures that could sustain liquid water, would eventually be able to give rise to life.

The third stage is retention of the atmospheric layer. Mars being smaller in mass than Earth however, has much less gravitational force and virtually no magnetosphere due to the cooling of its core and the loss of convection which generated the field long ago. Preventing the atmosphere being blown away by solar winds from the surface of the planet into Space presents a challenge.

In terms of the viability of terraforming technologies, there are additional obstacles. While we do currently have the know-how and the technological capability to radically change the planet's environment, by all accounts it would take at least 40,000 years (and that's based on optimistic estimates) to transform the surface of Mars to make it fit for human habitation. The high levels of carbon dioxide in the thin Martian air are ideal for triggering a runaway greenhouse effect; the problem however, is the production of enough oxygen to make the air breathable for humans. Not only would oxygen be difficult to extract, but sufficient levels of carbon dioxide would also need to be buried in order for the atmosphere to maintain its oxygen content. At the current rate at which CO_2 is buried by geological forces here on Earth—0.1%—it would take four million years to create an oxygen atmosphere fit for humans. Even if that process could be compressed by technologies to 10% of that time, it would still take 40,000 years.[3]

[3] James Kasting from the *Terraforming debate: Giving Mars back its heartbeat.*

Geoengineering to make the Martian environment suitable for plant life (a process known as ecopoiesis) would be much easier given the abundance of CO_2 there. Once a plant ecology is established, more oxygen would be produced as well, although still not enough to support human life. But no one in the Space industry seems eager to go the expense and effort of modifying Mars to support only plant life. What could be achieved in a more realistic time span is converting the atmosphere to make it partially habitable for human occupation. The runaway greenhouse effect could not only significantly warm up the planet's surface, but also create enough atmospheric pressure that humans would no longer require pressurized and heavily insulated spacesuits outdoors. Only breathing apparatus would be required.

4.4.4 Robots, Androids, and Artilects

Robots, like most technologies of utility, were originally designed as objects of service, relegated to tasks that are considered too dull, too dirty, too delicate or too dangerous for human beings to undertake themselves. What differentiates robots from other machines however, is their degree of autonomy. They can carry out their assignments with little supervision or oversight from its programmers. And as robots become more complex, they will not only be designed with even higher degrees of independence, but will also perform a wider range of functions. Besides providing basic service functions, social robots are being developed for companionship roles, in which they may eventually become caretakers, guardians, and even intimate partners. As they become more integral to our lives at a variety of levels, our relationships with them will change as well as our relationships to each other (Weiss et al. 2011).

4.4.4.1 The Use of Robots in Space

Robots are already an integral part of Space exploration, and though the public doesn't think of them as robots, the series of rovers and landers (Sojourner, Spirit, Pathfinder, Opportunity, Curiosity) are in fact, the first expeditioners to Mars. "Space probes hurtling through the solar system may not seem like robots, but they fully merit that name by performing programmed tasks over long periods without direct human supervision" (Hubbard 2005, p. 651). With the knowledge we have acquired through those missions, robots have blazed the trail for the future generation of human travelers. As we continue to venture farther into Space, they are expected to play an even larger role. And as humans migrate and eventually set up outposts in extraterrestrial locales, robots will become indispensable members of those communities.

There are numerous advantages of sending robots over humans to Space:

- Robots can better withstand the environmental conditions of Space: they can 'survive' extremely cold temperatures, exposure to radiation, and they don't need breathable air, pressurization, or gravity. Their one vulnerability is the fine lunar and Martian dust, an environmental hazard which can impede their proper functioning.
- Robots can obtain highly detailed information about the environment that would elude detection by human senses. "A robot can easily have radar, infrared, and gamma-ray eye, and so in principle a robot can see far more than a human can" (Landis 2003, p. 855).
- Robots do not need food, shelter, clothing, or other supplies, and they produce no waste.
- They are not bothered by isolation and confinement, or other psychosocial stressors.
- There is no need for them to return to Earth, but even if they did, physiological and psychosocial re-acclimation is not an issue.
- Robots are arguably more reliable than humans; they are not subject to illness, disease, fatigue (although they can get damaged and also require some degree of maintenance).

NASA has a number of robotic prototypes currently in development that will create new technological interfaces between humans and the Space environment (Hubbard 2005):

- An interactive model based on collaborative control, where the human and the robot are in dialogue with each other to accomplish an objective. In this model the robot is no longer treated as a subordinate entity, but has equal say and input in the problem-solving process, involving reciprocal exchange of information (Hubbard 2005). This direction for robotic development is based on a new paradigm called 'human-centered computing' (HCC), where the design of the systems is driven by the needs of the task—not by the complexity of the technology. This approach is very different than its predecessors in that rather than humans having to adapt to the machines, the machines are adapted to the humans.
- Other robotics interfaces involve the use of virtual reality technologies: VIZ creates a virtual world, utilizing 3D modeling of the planetary environment in which movements can be simulated and tested before commands are issued to the field robot. "VIZ was used throughout the MER mission to put teams working at mission control in Pasadena into the virtual Martian environment" (Hubbard 2005, p. 654).
- Machine Vision offers a form of telerobotic sight through the creation of immersive environments that enable humans to better assess situational conditions and analyze images of remote locations.
- An advanced anthropomorphic system called Robonaut was designed to achieve a high level of dexterity in order to use complex and multiple tools. The anthropomorphic design developed by engineers was not with the intention of

making the robots *appear* more human-like, but as a matter of practicality, to *function* in a more human-like way. Anthropometrics informs ergonomic design, and if robots are built with similar movements and proportions to humans, they can then use the same tools and be programmed to perform tasks much the same way that humans do. (Anthropomorphic features however, will come to serve another purpose when considered for the design of social robots.)

Because of the inherent risks and dangers of going to Space, there are many at NASA who believe that all future physical exploration of Space should be done exclusively by robots, and not by humans. NASA scientist Geoffrey Landis argues that sending robots effectively eliminates risk, and therefore more dangerous exploration missions can be undertaken. A Mars vehicle that did not have to be human-rated could also be developed faster and cheaper, using simpler technologies. And without any direct human contact with the planet, any indigenous Martian life would not be exposed to the threat of human contamination, nor would humans run the risk of becoming contaminated by any possible microbes on Mars; this in turn would eliminate the need for subsequent quarantine on Earth.

Finally, with regard to the experience of Space, Landis believes that telerobotics would have "all the excitement of being there, at a fraction of the price" (Landis 2003, p. 858). "With virtual reality, we can engage the entire population of Earth—everybody can be an explorer!" (Landis 2003, p. 860). While telepresence on Mars via virtual technologies would certainly be preferable to any of the reality shows that are currently popular on TV, would it really be as exciting as being there in person? There are many other aspects which make for richer experience that VTs cannot provide. The protracted time durations and the hardships associated with traveling long distances, the element of danger and risk, and the camaraderie of a shared physical adventure are factors that intensify the overall experience, which VT simulations do not provide—but perhaps that will change in the future as virtual technologies continue to develop.

All robotic technologies geared for Space exploration have until now focused on scientific purposes; little attention has been given to the possibility of utilizing them for social purposes. Given the small size of the crews that will pioneer Space, the remote destinations and their distances from other humans on Earth, and the durations of time required to travel to those locations, the presence of social robots can serve to mitigate problems of isolation, loneliness, and lack of mental stimulation in addition to the tasks for which they are currently designated for. They should be acknowledged as integral members of the crew rather than technological devices for mere service functions, and their social role will require assessment of ethical considerations, because "ethical relations between humans and artilects may indeed be vital for the success of future space missions. Long-range spaceflight missions that are planned and carried out without sufficient consideration of the ethical and emotional relationships between robots and humans could be doomed to failure, or worse" (Dator 2012, p. 61). As technologies continue to be developed for various aspects of Space exploration, the social role of robots should be given more serious consideration.

4.4.4.2 Sociability

In addition to service and scientific exploration, robots are also being groomed for social purposes, which present a new set of challenges and issues. "Rich affective interchanges will become increasingly important as robots begin to enter long-term relationships with people" (Breazeal 2001, p. 582). Assuming roles as caretakers and companions, robots will need to attain at least a minimal level of competence in social interaction. Social robots need not simulate the appearance of human beings, but must be able to exhibit the appropriate expressions and behaviors in social settings. When robots are capable of conveying even a limited range of emotional responses, humans are more likely to engage with them, despite their outward appearance (Walters et al. 2008; Breazeal and Scassellati 2002). This was depicted in the film *Moon*, in which a robot was the sole companion of the lone employee of a mining corporation stationed there. Though a rather clunky gadget that possessed no anthropomorphic qualities, it was able however, to effectively communicate its basic emotional states simply with the use of smiley-face emoticons (Fig. 4.4).

Recent studies have shown that humans can develop attachments to robotic pets much like they do to living ones (Weiss et al. 2009), and robotic pets have also proven beneficial in a therapeutic capacity. Though the mechanized objects were perceived as simulation of real, living and breathing creatures, the associated feelings of affection did not appear to be contingent upon an aspect of ontological authenticity. What was important in evoking an emotional response from the participant was that the robot be perceived as *behaving* sociably—that it displayed expressions that are consistent with the communicative exchange (Fig. 4.5). Even though children in studies intellectually differentiated the robotic dog AIBO from a flesh-and-blood dog, their behavioral interactions were similar with that of a real dog. Feelings of attachment and affectionate behavior arose in particular to behaviors that called for nurturing and attention (Weiss et al. 2009). The fact that people often demonstrate social behaviors towards inanimate objects, like cars and computers (Turkle 2007), is evidence of a predisposition towards sociability, regardless of the object involved in the interaction. Moreover, in observations of interaction with the robot Kismet, the manner in which the robot reacted to statements made by the human in turn elicited emotional responses from its human subject, creating a feedback cycle known as affective mirroring (Breazeal 2001).

The conclusions from this body of social research are a harbinger of the new kinds of relationships we will have with our intelligent inventions in the future. As the technology for developing greater sociability advances, humans will form deeper and more connected relationships to their robotic creations. Levy claims that sexual relationships and falling in love with our robots are inevitable—and even suggests that robots could make suitable marriage partners. For him, it is the feelings that the objects evoke that matter, not the status of the objects themselves: "If the alternative is that you are lonely and sad and miserable, is it not better to find a robot that claims to love you and acts like it loves you? Does it really matter [that

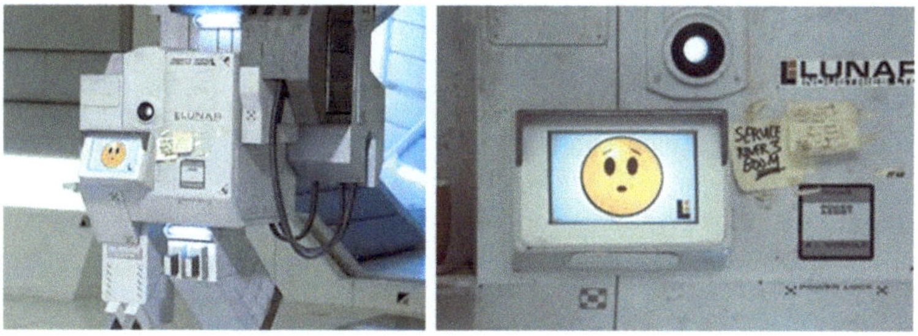

Fig. 4.4 Gerty, the empathetic robot from *Moon*

Fig. 4.5 Sociable Robots: AIBO the dog, Paro the seal, and Kismet

the object of your affection is a robot], if you're a happier person?. . .It just matters what you experience and perceive" (Levy 2009).[4]

But Turkle disagrees with this sentiment: relationships with machines "offer [the illusion of] companionship without demands of friendship to those who are lonely yet afraid of [human] intimacy. One could be a loner yet never alone" (Turkle 2005, p. 296). In lacking "the mutuality and complexity of a human relationship, machines seduce because they provide a chance to be in complete control" (Turkle 2005, p. 24). This may very well be true—how fulfilling can the illusion of unreciprocated love be? Robots however, could be programmed to have emotional needs of their own that must be satisfied, so that any relationship would be one of both give and take. The same argument has been made with regards to people who prefer the company of animals over humans, and in this context her objections seem less valid. Nevertheless, for those who have no better option, such as those living in conditions of isolation and confinement, companion robots would be a welcome alternative to utter solitude and would clearly enrich their experience if for nothing else, by providing stimulation, engagement, and emotional fulfillment. Adaptation has an emotional component; Breazeal writes that "Emotions. . .[are] centrally involved in determining the behavioral reaction to environmental (often social)

[4] From Levy's presentation of his book, *Love and Sex with Robots*, at the Museum of Sex in New York City.

and internal events...endowed by evolution because of their proven ability to facilitate adaptive responses to the vast array of demands and opportunities a creature faces in its daily life" (Breazeal 2001, p. 583). Therefore, anything that can facilitate a greater breadth and depth of emotional exchanges will be beneficial to the isolated crews who will settle on the Moon or Mars.

The nature of human interaction with robots depends in large part on their outward appearance. "Social interaction with people inherently requires the employment of a degree of anthropomorphic, or human-like, qualities whether in form or behavior or both" (Duffy 2003, p. 178). "Mechanical-looking robots tend to be treated less politely than robots with a more human-like appearance. Also, humans commonly treat mechanical-looking robots in a subservient way (i.e. less socially interactive) compared to more human-looking robots. Moreover, expectations are in general lower with regard to abilities and reliability for mechanical-looking robots" (Walters et al. 2008, p. 160). On the other hand, designing robots to be too human-like may also skew expectations in the other direction. Although "empathic androids...require something resembling a human body and a human face, in whose gaze we experience validation...This will not be achieved by the aggregation of ever more realistic behaviors and appearances (Vinayagamoorthy et al. 2005). Such technology creates, in the limit, only the perfect simulation" (Lee 2006, p. 426).

The goal of anthropomorphic design should not be to create an exact human replica, or even to try to convince people that the robot has similar capabilities of a human, but rather to strike a balance between peoples' expectations based on appearance and the robot's actual capabilities (Duffy 2003). If the robot bears a strong human resemblance but its behaviors don't live up to the same standard, this will cause dissonance in the social relationship.

Moreover, when made to look like synthetic humans, other problems arise. Ironically, when robot appearance became *almost* indistinguishable from humans, people experienced a phenomenon referred to as the Uncanny Valley (Mori 1970). Studies in this area revealed that human response to robots' appearance maintained high comfort levels either when the robots were sufficiently different in appearance, or conversely when they attained indistinguishable human likeness. However, when the robot's features fell into a zone somewhere in between, approaching likeness but belied by their robotic behavior or some other perceived defect, levels of discomfort rose markedly. The explanation for this is that anomalies in appearance are perceived as *deviations in human likeness* rather than *differences from human likeness*. The robot was construed as being no longer 'true' to its robotness, but rather attempting to be something inauthentic—in this case, human—thus creating an untenable tension between affiliations of 'us-ness' and 'them-ness' that resided more in the liminal area of 'not-us-ness'. This reaction "suggests a primitive distrust (or disgust) for fake persons. Such persons threaten our well-being, and we may have specific alarm responses to repel them" (Lee 2006, p. 426).

Duffy asks: "Can the *illusion* of life and intelligence emerge through simply engaging people in social interaction? How much can this illusion emerge through people's tendency to project intelligence and anthropomorphise?" (Duffy 2003,

p. 178). The language Duffy uses here presupposes that projection is equated with illusion, and does not leave room for the possibility that intelligence may arise from the very act of projection itself; that intelligence is fundamentally an autopoietic process. If we treat an entity as if it were intelligent, it is possible (or at least worth pondering) that our treatment is precisely the antecedent condition for intelligence to take hold and flourish, that we can by our interactions *imbue* intelligence. We have seen evidence that intelligence emerges through the relationships we forge with other entities, such as animals; it does not have to be defined as emanating from, or possessed by, a single solitary source. (In fact, theories of symbiogenesis, as well as of posthumanism, posit that intelligence is a shared, collective attribute, and not exclusive to the individual.)

Ultimately, the question of robotic intelligence is made irrelevant by the phenomenological perspective of consciousness. Our relations with robots need not be based on whether robots are intelligent, sentient, or conscious; there is no need to debate the validity of strong formulations of AI (which argue that human intelligence can be duplicated) and weak formulations (which claim that human intelligence can only be simulated). Their ontological status is not as important as our perceptions and experience through our interactions with them—thus eliminating the need for the problematic Turing test to determine whether they are human or computer.

Coeckelbergh similarly rejects the ontological argument in favor of framing the debate from the standpoint of human-technology relations. "The robot is neither part of me (embodiment relation) nor something that mainly mediates my relation to the world. Instead, in our interaction with the robot 'it' appears to us as more than a thing: it is an other to which I relate" (Coeckelbergh 2011, p. 198). He maintains that this alterity relation takes precedence over any ontological differences. This position—based on relations rather than on definition of being—will become the crux in defense of treating robots as ethical entities.

4.5 Alternative Approaches for Adaptation in Space

Technologies for highly specific functions in Space habitats should not be evaluated only by the primary level of activity associated with its use (i.e., the function for which it is intended), but also by what the foreseeable secondary (meta-activity, phenomenological) effects will be on adaptation to the environment. Teleoperative devices, for example, such as robotics and remote sensing, will be essential to the crew in order to interface with the immediate physical environment, but how these implementations will impact the processes of adaptation must not be overlooked. The second level of activity—the unforeseen social ramifications beyond its prescribed or intended purpose—must be examined in terms of how it changes the perceptions, attitudes, and behaviors of the individual users, as well as how it affects social interactions among the crew.

I have developed four archetypal approaches which span a range of technological intervention: Avoidance, Domination, Subordination, and Integration

Avoidance Strategy	Domination Strategy	Subordination Strategy	Integration Strategy
Humans | Nature (ex: sealed container; isolation & confinement)	Humans ↓ Nature (ex: terraforming)	Nature ↓ Humans (ex: genetic engineering)	Humans ↔ Nature (ex: symbiotization and cooperation)

Fig. 4.6 Alternative approaches for technological intervention

(Fig. 4.6). Avoidance is based on keeping humans separate from the natural environment; domination is based on humans controlling the environment; subordination is based on humans acquiescing to Nature; and integration is based on negotiating with the environment. Unlike archetypal scenarios which classify outcomes, these alternatives hinge on technological intention. They are not dependent on the viability of any specific technology. Since life-supporting as well as many discretionary types of technologies will be an essential part of our interaction with the Space environment, the question is: which type of intervention will best support long-term adaptation and what will be the implications for the relationship between humans and the Space environment?

In addition to these questions, there are also ethical and moral concerns raised in determining how technologies should be applied, as well as the speed with which they should be advanced. These issues are made more complicated by the fact that the Space environment presents a unique set of conditions and challenges that are without precedent. The justification for the direction we take may be a reflection of our existing worldview, or one that we aspire to.

These categories are not as clear-cut and absolute as I have defined them here. They can be present to varying degrees, and in fact each may have characteristics of other modes as well. They may even change according to the primary and secondary levels of activity associated with the use of technologies. But by defining categories concretely, it elucidates the range of possible approaches (Fig. 4.7).

4.5.1 Avoidance (of the Environment)

The avoidance strategy entails technologies designed with the express purpose to erect barriers—both material and phenomenological—in order to minimize or eliminate interaction with the environment, usually with the specific intent to unilaterally safeguard one element from exposure/vulnerability to another. One perspective is that humans need protection from an environment that is hostile and threatening. This is part of what shelters do: they protect its occupants from the harshness of exterior conditions (such as extreme weather) or other external dangers (predators). Taking defensive or preventative measures is programmed into our instinct for survival, and this phenomenon is ubiquitous in Nature: many organisms

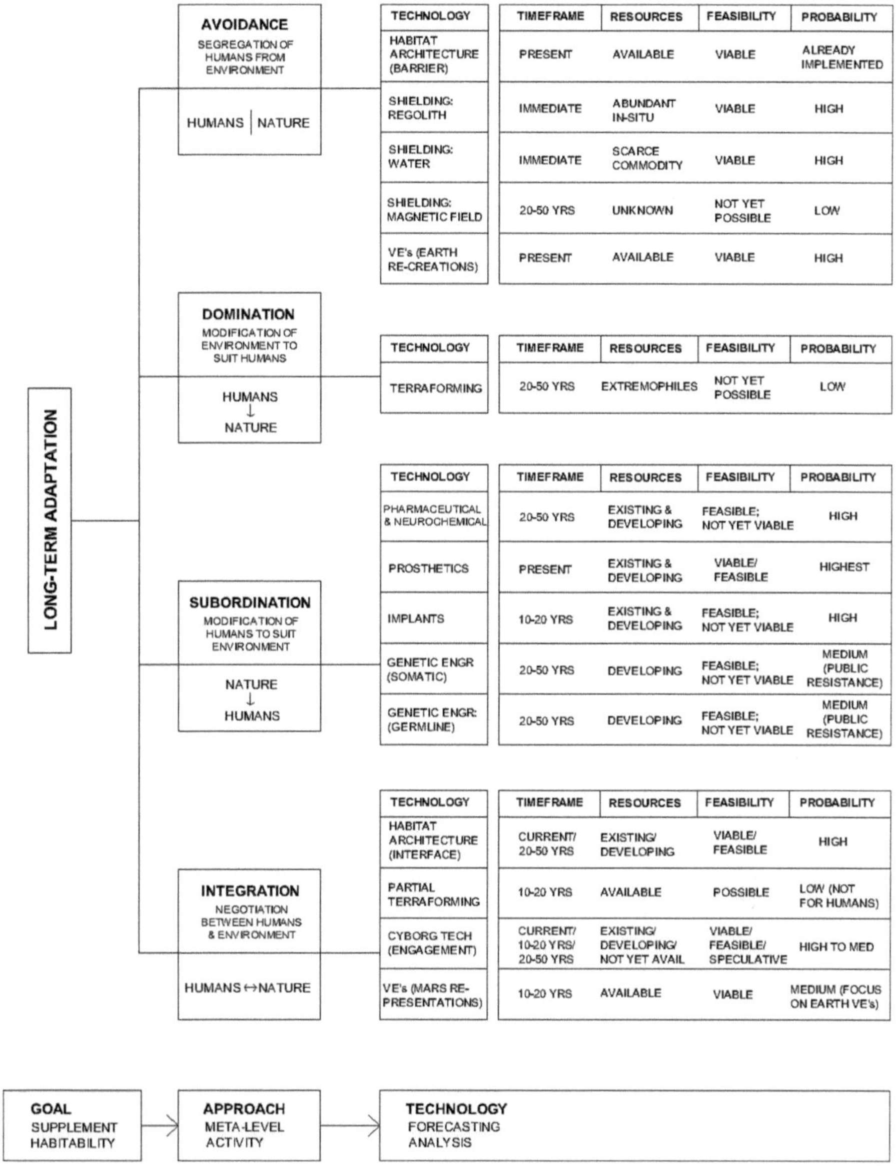

Fig. 4.7 Morphological analysis diagram

such as crabs, mollusks, and snails have all developed morphological sheltering mechanisms to minimize risk of harm or injury and protect themselves in dangerous encounters, as well as to afford the organism a strategic vantage point to evaluate further action. Such enclosures are not absolutely impenetrable; there must be a degree of permeability that allows for the transmission of information, energy intake, and waste output, so absolute isolation is not possible (Benyus 2002).

Similarly, the Space habitat has been conceived primarily as a protective barrier, a tightly sealed cocoon that prevents its occupants from exposure to any lethal physical conditions of Space. This has been historically the driving factor for its design, with functional requirements for performance a secondary criterion. Regolith or water shielding against cosmic radiation and solar flares is a clear example of a technology of avoidance. Despite their common functional objective of shielding however, there are qualitative phenomenological differences between the two that are worth noting. Water is fluid and translucent—thereby allowing movement within it and light to penetrate—and could conceivably support aquatic life; whereas regolith is sedentary and opaque, which would sever any unmediated visual connection to anything beyond the confines of the structure. The way in which these two types of shielding may be experienced by the habitat's occupants can either enhance or erode the perceptions of the external environment.

The reverse perspective is that the shelter serves as protection for the extraterrestrial environment against the risks and dangers of human occupation. The implicit assumption in this view is that humans are an invasive and contaminative presence that will disturb or mar the pristineness of the environment. Unlike the first perspective, the interests of the environment take precedence over those of the human, reflecting the cosmic preservationist's position that humans do not belong in Space and should not do anything that might alter it in any way.

Independent of the reasons that drive the avoidance approach, there are negative implications from a phenomenological perspective. When an experiential engagement with the outdoor environment is suppressed, crew members are also likely to feel like passive observers, with no domain of control beyond their internal world. A habitat that doesn't attempt to establish any kind of sensory connections to the exterior will only exacerbate feelings of confinement and isolation, and crew members will suffer from solipsistic tendencies due to lack of external stimuli and interaction. Moreover, the disjunctive nature of the intervention by the habitat also stands as a symbol of their status as intruder, not as affiliate. In the Avoidance approach, humans possess the least degree of agency, and in prohibiting confrontation to, and negotiation with, external environmental resistances, this approach represents the antithesis of the Integration strategy.

There are nonetheless, some advantages to this approach: it is the least technologically invasive in that it does not entail any genetic or biotechnical modifications of the body nor any modifications to the environment, it minimally impacts the Martian environment so that it can remain for the most part intact, and the habitat can recreate some of the comforts and familiarities of home on Earth. In the long-term however, an avoidance approach will grossly hinder the process of adaptation, and also raises the question of how long this model can be sustained. "Artificial atmospheres encapsulated in some sort of enclosure constitute only temporizing, and dangerously temporizing at that, since we put ourselves in the same position as a fish taking a small quantity of water along with it to live on land. The bubble all too easily bursts" (Clynes and Kline 1960, p. 30).

This strategy has defined the status quo of Space exploration and its assumptions have never been challenged; hence no alternate strategy has been given serious consideration. For now it seems to be the most practical for our current needs, and therefore it is highly probable that the proposals for settlements in Space will continue on this trajectory. The technologies already exist and have been proven effective, so there is little or no incentive to change direction, especially when it entails greater risks to the crew, greater economic investment, and no tangible benefits—scientific, economic, or otherwise. But in the long run from the point of view of adaptation, it will simply become untenable.

4.5.2 Domination (over *the Environment)*

This approach to the environment prescribes technologies that are designed with the intent to control or manipulate the habitat and/or physical surroundings exclusively for human advantage. In this anthropocentric approach that places the interests of humans above all else, technological intervention customizes the environment to the human organism instead of the human organism itself having to adapt to the environment. On Earth, we build dams, construct retaining walls, blast bedrock, and dredge shoreline in our attempt to subjugate the 'obstacles' of the environment and shape it to serve human goals. This approach is in some ways similar to the avoidance strategy insofar as there are no attempts at negotiation; the actions are unilateral.

As the status quo approach for how we utilize technologies on Earth, most of our technologies since the Scientific Revolution have been directed towards subduing the 'wilderness' of Nature—an extension of the humanist principles that came out of Newtonian/Cartesian thought, and later reinforced by Darwinist theory. This attitude towards Nature is founded on the assumption that Nature is feral, non-rational, and at the disposal of human beings for their exploitation. Moreover, human intervention is needed to tame Nature and bring it under its reign. Thus the status of humans is elevated above that of the environment. All other entities have less standing and less agency, and are therefore expendable if our interests conflict with theirs. As in the avoidance and subordination scenarios, we find ourselves outside the realm of Nature—at odds with it and in a state of eternal conflict—not a participant in its activities and processes.

Should technologies of domination be used to 'domesticate' the Martian environment in order to accommodate human life, much like we have done on Earth? On Earth we tend to supplant the organic with the inorganic; we clear away the wilderness to make way for buildings. On Mars it would be reversed: we would replace the inorganic with the organic, seeding life where none exists. The option of terraforming Mars, whose environment possesses potential to be transformed into an Earth-like atmosphere, is a subject of intense debate. There is very little consensus from either a scientific or ethical perspective on several critical matters: First and foremost, does any life exist there or not? The evidence that indicates the presence of vast amounts of water-ice below the permafrost and the scant amount of methane found in the atmosphere, as well as the recent discoveries of extremophiles on Earth living in conditions originally thought to be far outside the range that could support any life, certainly makes the prospects for finding microbial life more promising. Yet scientists are still split on this issue; some believe it is very likely, while others believe the chances are slim. With continued rigorous explorations of the planet, we may definitively know sooner than later, but the likelihood of finding some form of prokaryotic organism seems to increase the more we discover about the conditions on Mars.

In the wake of such a possibility, the debate over whether we should attempt to terraform Mars spans all the various positions: On the pro-terraforming side, justified by an anthropocentric rationale, are aeronautical engineer and author Robert Zubrin, terraforming expert Martyn Fogg, astronomer Carl Sagan, and scientist Richard Taylor. There is no equivocation or contingency in their advocacy; modification of the Martian environment is a necessary step to further human interests in the solar system, and thus morally imperative. Adopting a more moderate biocentric stance in which terraforming should be pursued under certain conditions are NASA planetary scientist Christopher McKay, author Kim Stanley Robinson, astrobiologist David Grinspoon, and evolutionary ecologist John Rummel, who take the position that we should terraform only once we are completely assured that an alien planet does not harbor life of its own. But if it does, we should engineer the planet's environment to artificially nurture the alien life and help it thrive and co-evolve—or even co-exist with humans—but not try to reshape the planet solely for our own use.

Independent of the question of the existence of endogenous life, another area of dispute is whether humans should terraform at all, even if conditions prove favorable to do so. At the very least it would radically and irrevocably alters the planet's ecology. Assuming that exhaustive exploration can conclusively rule out the presence of even any microorganisms there, do we have the right to inseminate it with life from Earth, human or otherwise? Is the absence of indigenous life sufficient to justify that course of action? Again, there is no agreement on this issue. In 'The Terraforming Debates,' McKay and Grinspoon argue that since Mars once supported life (unlike the Moon which has always been thought to be dead), terraforming would be an *imperative* restoration effort. However, some environmental ethicists take a very different position in which the presence of life is not necessarily the determining factor. Those who are resolutely opposed to deliberate human intervention on Mars are geologist Lisa Pratt and science fiction writer Greg Bear, who argue for a preservationist approach in which all of Nature—whether it includes living organisms or not—has the right to exist as it is; in this view even inorganic matter has some degree of standing.

Despite the fundamental, and seemingly irreconcilable differences between the experts, all unequivocally agree that we simply do not know enough yet about Mars—or even about Earth, for that matter—to make a properly informed decision. What is known is that Mars once was a warmer, wet planet with an atmosphere, which could have very likely given rise to life at that time. Until any fossils are uncovered though, this remains conjecture. And almost no one advocates terraforming if even the simplest form of life exists until we are sure that whatever we introduce to the environment is not toxic to the extant organisms there, or that we can effectively partition the local ecologies so they are not affected. Zubrin is the dissenting exception; he endorses the more extreme position that it is an ethical imperative to terraform Mars, regardless of whether microorganisms exist there already or not. "[His] proposal would be implemented before even knowing what those Martian microbes might offer, what they might be able to teach us, and prior to any consideration of the complexity of their ecosystems or their roles in maintaining it, much less any ethical consideration for their existence" (Kramer 2012, p. 123). If any life is found however—especially a second genesis—this discovery would be one of the most enlightening and profound events in furthering our understanding of the Universe. It would radically transform our cosmological view: we would no longer be alone in the vast expanse of Space, and discovery of life elsewhere would seem more likely.

But besides the scientists, environmentalists, and engineers, the lay public too (though for different reasons) has serious reservations about terraforming which could undermine any political will to successfully undertake such an ambitious endeavor—even *if* all the experts eventually came to agree that terraforming is the correct thing to do. The ecological movement, sparked by books such as Rachel Carson's *Silent Spring*, Jared Diamond's *Collapse*, and *The Limits to Growth* by Meadows et al., has gained considerable momentum in the industrialized world over the last half century. As a result, there is more scrutiny of the consequences of our actions on the environment. The public is much more concerned now than at the beginning of the Space exploration program (when the U.S. was enjoying its height of post-war prosperity) about any damage or destruction we may wreak beyond Earth, given our recent history of what humans have done to our own global ecology through global warming and environmental pollution. This concern is often accompanied by economic considerations: why spend all this money to terraform Mars when we should be concentrating our efforts on repairing the Earth's ecology? These objections may vanish, however, should we come to believe that our extinction on Earth is imminent and migration to Space is our last resort for survival. Of all the possible strategies for human habitation on Mars, terraforming presents the least risk and is therefore most likely to insure our survival as a species.

Anthropologist Gregory Bateson presents an argument from an evolutionist's point of view that would defend terraforming as the most favorable option to pursue. He classifies organisms into three categories: adjusters, regulators, and extra-regulators. In 'adjusters,' environmental variables are allowed to enter the body and the organism must cope with its effects; in 'regulators,' homeostatic processes operate close to or at the input and output points of the individual organism; in contrast to both, 'extra-regulators' achieve homeostatic equilibrium by changing and controlling the environment, not the organism itself. By relocating control to outside the body, "natural selection, in the long run, favors regulators more than adjusters, and extra-regulators more than regulators. This seems to indicate that there is a long term evolutionary advantage to be gained by centrifugal shifts in the locus of control" (Bateson 1963, pp. 537–538). From this we can conclude that changing our environment on Mars to fall within the limits of human somatic flexibility would be more advantageous than other means of adaptation.

Another consideration is that technologies that would recreate a familiar Earth-like ecology would be far less threatening than those that would compel humans to undergo major modifications to their bodies which disturb our notions of humanity. If living on Mars came down to a life-and-death choice, terraforming would likely be more acceptable to people than radical genetic alteration. By shifting the burden of change to the planetary body rather than the human body, it is also the only option that would make circumstances possible for humans to adapt within their range of somatic flexibility.

There are other phenomenological implications that should be taken into consideration as well. If humans choose to radically modify and control the Mars environment, we will be adopting exactly the same strategy that we have done throughout our history on Earth. Although we may have learned from our original mistakes, merely adopting the process itself will still imply that human life takes precedence over everything else. In short, more value is placed on human existence and, by extension, other forms of life as well as the geology of the planet have less standing. Transforming the environment to suit human needs at best relegates any living organisms to passive entities and undermines human negotiative interaction with their habitat that I believe is essential to evolution.

But besides the negative phenomenological implications, the degree of viability at this time makes this approach moot. In theory we know how to change the conditions of Mars to both warm the planet and generate a denser atmosphere which would allow plant life to flourish, given the relative abundance of carbon dioxide already present in the Martian air. But to make the planet suitable for unmediated human habitation would require the production of significant amounts of oxygen, which given our current technologies would, by all accounts, optimistically take tens of thousands of years. This prohibitive factor makes total terraforming highly unlikely, but partial terraforming possible. However, given the ambivalence among experts in the field, and the lack of definitive evidence of the presence or absence of life on the planet, it seems that there would be no incentive to undertake this strategy at this time.

4.5.3 Subordination (to *the Environment)*

In this approach, technologies are designed with the intent not to change the environment, but to accommodate it by adjusting to *its* vicissitudes. The exclusive locus of intervention is the subject itself. The nature of the intervention is to artificially augment the somatic flexibility of the organism. A basic example of this is clothing, a prosthetic technology which supplements the biological process of regulatory adjustment. A more advanced example is genetic enhancement which alters the body so that it can directly adapt to the surroundings without prosthetics or other physically mediative technologies. In its extreme, subordination technologies eliminate any form of resistance and thus negotiation between subject and environment is rendered superfluous. Its intent, similar to that of the avoidance and domination scenarios, is also non-integrative by its nature.

The process of altering humans and other life forms (through genetic engineering or other biotechnological means) in order to make them better suited to inhabit Space environments is known as pantropy (Morgan 1980, p. 147). Technologies aimed at modifying the subject for the purposes of survival would have to be far more intensive in Space than those used on Earth. With minimal or no intervention humans are able to adapt to less than optimal conditions on Earth—such as lower levels of oxygen and reduced pressure at higher altitudes—but only within certain regulatory and acclimatory limits. However, the conditions of Mars—the trace amount of oxygen and the 0.1% atmospheric pressure—would far exceed our homeostatic capabilities. Without some kind of mediation, the human body could not adjust to these conditions, and would simply die. Technologies would have to pick up the considerable slack in the difference between the physical conditions of the environment and our biological capacities for adjustment. Several types of technologies could be employed for the purpose of radically altering the human body either genetically, chemically, or morphologically.

Pharmaceuticals, somatic gene therapies, and nanotechnologies affect only the individual, and because they are not heritable, die with the subject. Unlike somatic gene therapy however, manipulations done through germline therapy—genetic engineering that targets the reproductive cells—would however, be heritable and any changes would be passed on to succeeding generations. This course of technological intervention is what Clynes and Kline proposed when they first conceived of the cyborg for Space travel. The cyborg, as they defined it, was a human being that was modified with the purpose of being able to survive the conditions of Space. They favored this approach because, they argued, it was highly impractical to create closed ecological life support systems for sustaining human life in Space. The risks entailed are too great: should any critical component or subsystem of the habitat architecture fail, it would more than likely result in death for the entire crew aboard. Genetic engineering that would make the body suited to the idiosyncratic conditions of an extraterrestrial environment—though not yet viable in their time—was the only alternative that made sense because it gave the individual the capacity to adjust somatically on its own. "If [humans] attempt partial adaptation to Space conditions, instead of insisting on carrying [their] own environment along with [them], a number of new possibilities appear. One is then led to think about the incorporation of integral exogenous devices to bring about the biological changes which might be necessary in [human's] homeostatic mechanisms to allow [them] to live in Space qua natura" (Clynes and Kline 1960, p. 30). And if evolution favors only those traits that confer a fitness advantage, while rendering others obsolete, then genetic traits which allow us to thrive in a range of extraterrestrial conditions would eventually take hold and flourish.

Technological alteration or supplementation of the human body is the primary theme that is extensively explored in performance artist Stelarc's work and writings, and draws parallels to Clynes and Kline's position. He argues that "off the

Earth, the body's complexity, softness, and wetness would be difficult to sustain. The strategy should be to hollow, harden, and dehydrate the body to make it more durable and less vulnerable" (Stelarc 2000, p. 118). Stelarc believes that "the body is ill-equipped, not only in terms of its sheer survival, but also in its inability adequately to perceive and perform in the immensity of outer space. Rather than developing specialist bodies for specific sites, we should consider a pan-planetary physiology that is durable, flexible, and capable of functioning in varying atmospheric conditions, gravitational pressures, and electromagnetic fields" (Stelarc 2000, p. 120).

Another type of intervention would be prostheses and implants which change the body morphologically but not genetically. Prosthetic devices could be developed specifically to allow humans greater exposure to the Space environment. Such inventions already exist in the form of pressurized and thermally-insulated spacesuits. Sensory augmentation devices of the future—whether they be prosthetics, implants, pharmaceuticals, or genetic engineering—could conceivably allow sight beyond the visible spectrum, hearing that exceeds the normal range of audibility, and smell that can convey detailed information about our immediate environment.[5]

The proposition of the human individual carrying the burden for change, is the most likely to meet with resistance, fear, and hysteria—reactions that would only be exacerbated by the speed and magnitude with which change could be introduced. Technologies of subordination will no doubt threaten our understanding of what it means to be human, no matter which type of technology is applied. This is evident in the proliferation of dystopian images of the nonhuman/inhuman in science fiction, as well as in the burgeoning literary discourse on cyborgism and posthumanism.

There are other implications of the subordination strategy that merit examination. The implicit assumption underlying this approach is that human presence essentially contaminates Nature, particularly through our technologies. Therefore the human subject must bear the burden of accommodation so that the environment may be preserved in its existing condition. This attitude is similarly reflected in the current sustainability movement, which advocates minimizing the human footprint, reducing consumption and waste, conserving resources, preserving Nature's pristineness, and even removing human activity of any kind—prescriptions that no other form of life on Earth adheres to. In this view, humans are seen as something separate, distinct, and fundamentally destructive to Nature. Humans—and by extension our technologies—are therefore unnatural. Preservation of Nature takes precedence over human life; thus the 'interests' of the natural environment (however those may be defined) stand above those of the human.

[5] If we move away from the narrower and more conventional definition of prosthesis as a replacement for a lost appendage to a more encompassing definition as augmentation of bodily capacities, then all technologies would effectively fall under this definition.

Although the most practical approach in theory, given the controversy that engulfs genetic-altering technologies, along with the fact that no such technologies are foreseeably viable, and the lack of political will to move it forward, the probability of adopting this approach is relatively small compared to other options. Genetic engineering is still in its nascence and because it is so mired in debate over ethics, the research is slow to gain momentum and any breakthroughs often are accompanied by social outcry. Public resistance aside, the development of a superhuman capable of withstanding the radically different physical conditions of Space has not been a research priority in the Space industry in the way that robotics has. Citing both the dangers and the complications of sending humans to Mars, NASA Ames director Pete Worden advocates sending technical surrogates to explore the planet in lieu of humans.

*4.5.4 Integration (*with *the Environment)*

We are "children of the Universe…not only of Earth…"—Ray Bradbury, *The Martian Chronicles*

A fourth option to be considered is the use of technology to develop relationships of integration between humans and the Space environment, with the intent of promoting reciprocity so that all entities can be mutually beneficial to each other and can prosper. This cooperative approach is derived from Lovelock's Gaian notion that subject and environment are not distinct entities, but rather co-evolving elements of a cohesive homeostatic feedback system. Extrapolating from the Gaian perspective to the cosmological scale, human presence does not 'contaminate' extraterrestrial environments, even should it be the case that indigenous life forms on Mars exist. But the prevailing attitude (at least in the industrialized West) is that humans are interlopers in Nature—and as outsiders, we observe, we explore, and we exploit, but we do not belong.

An integrative approach towards adaptation to the Space environment presumes that humans are not separate from Nature, and recognizes the essential interdependence between all living things and their environments. Biologist Lynn Margulis observes that "the tendency of independent life is to bind together and reemerge in a new wholeness at a higher, larger level of organization," stipulating that "the near future of *Homo sapiens* as a species requires our reorientation towards the fusions and mergers of the planetmates that have preceded us in the microcosm" (Margulis 1998, p. 12). She asserts that "living together will be as crucial to the colonization of outer space as symbiosis and diversity were to the Paleozoic Era colonization of dry land. Life in Space, if it is to occur, will require physical alliances, including new symbioses, among differing life-forms" (Margulis 1998, p. 106). Though her prescription may sound like an argument in favor of terraforming, what Margulis and Lovelock propose does not advocate putting human interests above the interests of other entities, but rather on par with them.

This means that we shouldn't necessarily avoid terraforming, but that the emphasis should be on fusions and mergers, not on conquest or control.

Symbiosis between humans and other life forms is but one type of integrative mediation that enhances viability for life in Space. Symbiotic integration can also occur not only between humans and other living organisms in Space, but also between living organisms and their physical environment, and even between humans and their products—a view that is put forth by Stelarc when he writes:

> "The problem with space travel is no longer the precision and reliability of technology but with the vulnerability and durability of the human body...The issue is how to maintain human performance over extended periods of time. Symbiotic systems seem the best strategy: implanted components can energize and amplify developments; exoskeletons can power the body; robotic structures can become hoists for a body insert" (Stelarc 2000, p. 121).

To maximize the potential for adaptation, integration can and should occur at all levels between disparate entities.

Symbiotic interaction is predicated on the condition of heterogeneity (Maruyama 1978, 1994). Because homogeneous organisms rely on the same resources, when those resources become scarce (or as in the case with humans, when they are valued or coveted) individuals are forced to compete with one another to acquire them in order to survive. Organisms that are similar have nothing unique to offer each other, and therefore there can be no basis for reciprocation; each organism is left to fend for itself. Dissimilar entities on the other hand, do not seek out the same resources, thus eliminating the need to compete. And by virtue of their dissimilarity, each individual may possess attributes that the other may benefit from, giving an incentive towards reciprocal behavior. Integration through symbiosis is premised on the *vitality* of heterogeneous entities (namely, the quality of being vital) rather than the *equality* of homogeneous entities. This shift eliminates the problematic comparisons through quantifiable measure that arises with notions of equality. Within homogeneous populations based on the premise of equality, each individual is viewed as a potential threat to the survival of the other; in heterogeneous populations based on the premise of vitality, each individual is indispensable to the survival of the other.

On a phenomenological level, integration strategies are more conducive to establishing connections and increasing familiarity through engagement with the environment. In the earliest stages of migration, the circle of empathy will be small, extending only to other crew members. As the crew becomes more familiar with their environment, expanding kinships can develop. In order for this to happen, there must be exposure to other entities that will then lead to negotiation. This confrontational process will also help alleviate feelings of isolation and confinement, as well as discourage solipsistic tendencies which are associated with ICEs by increasing awareness and familiarity with the exterior surroundings.

The integrative approach adopts the deep ecology view of humans as part of Nature, not as something distinct from or superior to it. However, it diverges from the preservationist perspective which proscribes any human intervention in Nature, implying that Nature is pristine and human presence is contaminative. For the preservationist, change should be 'natural'—but humans in fact belong to the environment as much as any other life form. All life is programmed through evolution to be motile through various means in order to perpetuate itself. Through evolutionary mobility, all species are invasive at some time or another, and therefore no species can claim to belong to a place more than another one does. It is simply a matter of which one arrived first. For all life, it is impossible to have presence without leaving a footprint—this is the fallacy of the preservationist philosophy. The question is not about whether to leave a footprint, but rather what kind of footprint we want to leave.

Environmental transformation is natural and inevitable—whether by human beings or other forms of life—and the processes of evolution are contingent upon conditions that fluctuate. The only difference between human-induced change and change brought about by other life forms resides in the aspect of intentionality. Humans, it is argued, can deliberately and willfully manipulate the environment through crafted, artificial means. But the use of technologies—thought to be antithetical to Nature—is not the exclusive domain of humans, rather it is an activity shared by all living things (Margulis 1997b, p. 36). Preservationism denigrates human life by characterizing it as contaminative. It cleaves us from Nature, and relegates Earth to a human ghetto by making the claim that Space is not their 'home.' Just as preservationists view humans as separate from Nature, they do not see Earth as part of Space, a decidedly non-integrative view. Fogg argues that the difference between bacteria who may happen to migrate to Mars (in a hypothesized process referred to as panspermia) and humans who migrate to Mars, lies in intentionality (Fogg 1998). But can we truly distinguish between accidental and intentional behaviors? If all life is innately driven to expand its niche in order to improve its chances for self-perpetuation, then it certainly follows that our drive to explore Space emanates from our own evolutionary instinct.

Integration through technologies of engagement will change our perception of the Space environment from one of hostility grounded in the notion of conflicting interests, to one of affinity based on the recognition of common interests. It is derived from, but also cultivates, respect for life and all of Nature, not just considering benefits to humans but also to the environment. The goal of striving for peaceful co-existence and mutually beneficial alliances doesn't permit unilateral exploitation. All technologies in the previous options were based on expedience; integration is the only approach that is based on experience.

There is no single technology that guarantees integration. All of the proposed technologies for Space discussed previously may be applied either to promote integration or to inhibit it. The habitat architecture can re-present the Martian landscape and reframe perceptions to facilitate connections and familiarity. Terraforming or other ecological modification could be done conservatively in a way that would support both indigenous life and introduce new forms of life. (The debate over whether to protect extant life on Mars or promote conditions suitable for human life is always framed as mutually exclusive interests, but the issue has never been discussed in terms of the possibility of supporting both in a cooperative manner.) Cyborg devices such as implants and prostheses could be selectively employed for the purposes of enhancing experience. Their use should be limited and discretion should be exercised; they shouldn't be used to make living in Space more expedient. Virtual reality can provide mediated experience of the Space surroundings that would otherwise not be possible. VRs that engage sight and hearing are already common; smell, taste, and touch—because they require direct contact with the body—will call for more sophisticated interventions. Sports equipment and accessories are the ultimate technologies of engagement because athletic endeavor is all about reveling in and overcoming resistances. The invention of games that arise from the unique conditions of the planetary environment will call for new technologies that may prove to be the most integrative of them all. The uses of various technologies are described in greater detail in Chapter 5 under cooperative strategies for technological adaptation.

It is difficult to assess the current capabilities as well as the probabilities for such a strategy because just how technologies may promote integration has more to do with intentionality of the user than with instrumental nature of any particular invention. What I see as the main obstacle for adopting this approach is the lack of understanding of the broader factors required for adaptation. The past focus of technological development for sending humans to Space has been primarily on safety and work performance; the importance of comfort and well-being have only recently been recognized and studied though not yet implemented. But issues surrounding long-term adaptation have not been acknowledged at all, much less taken into consideration in the design process. Questions of technological implementation cannot even be raised, until our relationships to non-Earth environments are acknowledged as a critical factor in adaptation.

Another obstacle to this approach is the degree of risk involved, and it may be difficult to demonstrate in quantifiable terms the actual practical benefits of enhanced experience that justify increased exposure and vulnerability to the potentially lethal elements of the environment. However, I believe this approach ultimately will best promote long-term adaptation. If at some point in the futures of human factors research the experiential aspects of adaptation are recognized, then this option may have a chance to be taken more seriously.

However, cooperation is ultimately necessary because the goal of human habitation in Space is too large, too complex, too technologically demanding, and too expensive an undertaking for a single Space agency or institution. "No one nation—no matter how wealthy or technologically advanced it might seem at the time—will be able to maintain the political commitment, entrepreneurial drive, and technological innovativeness to stay perpetually ahead in Space. . .for the future in Space we have to look beyond competition between terrestrial states" (Finney 1986, p. 206). New alliances for cooperative endeavors in the Space sector have already formed, not only internationally but also between the public and private sectors. It would have been impossible to imagine 30 years ago that the U.S. and the former Soviet Union would be collaborating with each other on Space-related endeavors—especially when the U.S.'s original motivation for venturing into Space was to assert American technological superiority. The shift from a competitive to a cooperative relationship is evidence that integrative processes, even at such a large scale and under antagonistic conditions, are not only robust, but can advance our goals more effectively than zero-sum interactions. Cooperation, which is discussed further in Chapter 5, provides the best option for long-term adaptation and species longevity in Space.

References

Basalla, G. (1989). *The evolution of technology*. Cambridge, MA: Cambridge University Press.

Bateson, G. (1963). The role of somatic change in evolution. *Evolution, 17*(4), 529.

Benyus, J. M. (2002). *Biomimicry: Innovation inspired by nature*. New York, NY: Harper Collins.

Biocca, F. (1997). The cyborg's dilemma: Progressive embodiment in virtual environments. *Journal of Computer-Mediated Communication, 3*(2). Retrieved from http://www.mindlab.org/images/d/DOC692.pdf

Breazeal, C. (2001). Affective interaction between humans and robots. In J. Kelemen & P. Sosík (Eds.), *Advances in artificial life* (pp. 582–591). Berlin, Germany: Springer.

Breazeal, C., & Scassellati, B. (2002). Robots that imitate humans. *Trends in Cognitive Sciences, 6* (11), 481–487.

Clynes, M., & Kline, N. (1960). Cyborgs and space. In C. H. Gray (Ed.), *The cyborg handbook* (pp. 29–33). New York, NY: Routledge.

Coeckelbergh, M. (2011). Humans, animals, and robots: A phenomenological approach to human-robot relations. *International Journal of Social Robotics, 3*(2), 197–204.

Danius, S. (2002). *The senses of modernism: Technology, perception, and aesthetics*. Ithaca, NY: Cornell University Press.

Dator, J. (2012). *Social foundations of human space exploration*. New York, NY: Springer.

Dimendberg, E. (1995). The will to motorization: Cinema, highways, and modernity. *October, 73*, 91–137.

Dodge, M., & Kitchin, R. (2001). *Mapping cyberspace*. London: Routledge.

Duffy, B. R. (2003). Anthropomorphism and the social robot. *Robotics and Autonomous Systems, 42*(3), 177–190.

Finney, B. (1986). The prince and the eunuch. In B. Finney & E. Jones (Eds.), *Interstellar migration and the human experience* (pp. 196–209). Berkeley, CA: University of California Press.

Fogg, M. J. (1998). Terraforming Mars: A review of current research. *Advances in Space Research, 22*(3), 415–420.

Fukuyama, F. (2002). *Our posthuman future: Consequences of the biotechnology revolution*. New York, NY: Farrar, Straus & Giroux.

Garreau, J. (2005). *Radical evolution*. New York, NY: Broadway Books.

Gjerde, K. (2000). Terraforming: The modern alchemy—or tomorrow's engineering technology? Retrieved from the Terraforming Information Pages at http://www.users.globalnet.co.uk/~mfogg/

Harries, K. (1998). *The ethical function of architecture*. Cambridge, MA: The MIT Press.

Hayles, N. K. (1999). *How we became posthuman: Virtual bodies in cybernetics, literature, and informatics* (1st ed.). Chicago, IL: University of Chicago Press.

Heim, M. (1998). *Virtual realism*. Oxford, England: Oxford University Press.

Hillis, K. (1999). *Digital sensations: Space, identity, and embodiment in virtual reality* (1st ed.). Minneapolis, MN: University of Minnesota Press.

Hubbard, S. G. (2005). Humans and robots: Hand in grip. *Acta Astronautica, 57*(2–8), 649–660. doi:10.1016/j.actaastro.2005.03.054.

Hughes, T. P. (2004). *Human-built world: How to think about technology and culture*. Chicago, IL: University of Chicago Press.

Kelly, K. (1994). *Out of control: The new biology of machines, social systems, and the economic world*. New York, NY: Basic Books.

Kramer, W. (2012). *Bioethical considerations and property rights issues associated with the discovery of extraterrestrial biological entities: Implications for political policy in the context of futures studies*. Doctoral dissertation, University of Hawai'i, Manoa.
Kurzweil, R. (2005). *The singularity is near: When humans transcend biology*. New York, NY: Penguin.
Landis, G. A. (2003). Robots and humans: Synergy in planetary exploration. *AIP Conference Proceedings, 654*(1), 853–860.
Lanier, J. (2010). *You are not a gadget*. New York, NY: Random House.
Lee, B. (2006). Empathy, androids and "authentic experience". *Connection Science, 18*(4), 419–428.
Levy, D. (2009). *Love and sex with robots*. New York, NY: HarperCollins e-books.
Mander, J. (1991). *In the absence of the sacred: The failure of technology and the survival of the Indian nations*. San Francisco, CA: Sierra Book Club.
Margulis, L., & Sagan, D. (1997b). *Microcosmos: Four billion years of microbial evolution* (1st ed.). Berkeley, CA: University of California Press.
Margulis, L. (1998). *Symbiotic planet: A new look at evolution*. New York, NY: Basic Books.
Maruyama, M. (1978). Symbiotization of cultural heterogeneity: Scientific, epistemological, and aesthetic bases. In M. Maruyama & A. Harkins (Eds.), *Cultures of the future*. Berlin, Germany: De Gruyter Mouton.
Maruyama, M. (1994). *Mindscapes in management: Use of individual differences in multicultural management*. Alder-shot, England: Dartmouth.
McKibben, B. (2003). *Enough: Staying human in an engineered age*. New York, NY: Holt & Co.
McLuhan, M., & Lapham, L. H. (1964). *Understanding media: The extensions of man*. New York, NY: McGraw-Hill Book.
Moravec, H. (1988). *Mind children: The future of robot and human intelligence*. Cambridge, MA: Harvard University Press.
Morgan, C. (1980). *Future man*. Devon, England: Irvington.
Mori, M. (1970). The uncanny valley. *Energy, 7*(4), 33–35.
Stallabrass, J. (1995). Empowering technology: The exploration of cyberspace. *New Left Review, 211*, 3–32.
Stelarc. (2000). From psycho-body to cyber-systems: Images as posthuman entities. In D. Bell & B. Kennedy (Eds.), *The cybercultures reader* (pp. 560–576). London: Routledge.
Stone, A. R. (1996). *The war of desire and technology at the close of the mechanical age*. Cambridge, MA: MIT Press.
Turkle, S. (2005). *The second self: Computers and the human spirit*. Cambridge, MA: MIT Press.
Turkle, S. (2007). *Evocative objects: Things we think with*. Cambridge, MA: MIT Press.
Vinayagamoorthy, V., Steed, A., & Slater, M. (July 2005). Building characters: Lessons drawn from virtual environments. In *Proceedings of Toward Social Mechanisms of Android Science: A CogSci 2005 Workshop in Stresa, Italy* (pp. 119–126).
Walters, M. L., Syrdal, D. S., Dautenhahn, K., Te Boekhorst, R., & Koay, K. L. (2008). Avoiding the uncanny valley: Robot appearance, personality and consistency of behavior in an attention-seeking home scenario for a robot companion. *Autonomous Robots, 24*(2), 159–178.
Weiss, A., Igelsböck, J., Wurhofer, D., & Tscheligi, M. (2011). Looking forward to a "robotic society"? *International Journal of Social Robotics, 3*(2), 111–123.
Weiss, A., Wurhofer, D., & Tscheligi, M. (2009). "I love this dog"—Children's emotional attachment to the robotic dog AIBO. *International Journal of Social Robotics, 1*(3), 243–248.
Winner, L. (1986). *The whale and the reactor: A search for limits in an age of high technology*. Chicago, IL: University of Chicago Press.
Winner, L. (1989). *Autonomous technology: Technics-out-of-control as a theme in political thought*. Cambridge, MA: MIT Press.

Chapter 5
The Case for an Integrative Approach

5.1 Premise

Because integration relies on reciprocity, cooperation must be the basis for any integrative approach. Having proven to be evolutionarily stable, cooperative structures promise greater collective gains than rivalries do, they lead to more robust and durable relationships, and they better promote long-term adaptation. In short, cooperative strategies for Space exploration and habitation will better insure the success of extraterrestrial settlements than competition-based strategies that currently define the status quo.

The driving force behind human exploration of Space—reflecting the prevailing Darwinian worldview—has been one of conquest and exploitation: the military wants to control Space for defense purposes; entrepreneurs seek economic opportunity in mining planets for its resources or in creating tourist destinations in Space; nations desire to extend their sovereignty to new territories in the quest for superiority and power; and even the most benign intentions to 'colonize' Space are redolent of imperialist motives and manifest destiny. If we continue on this same path, we are destined to the same fates that we have brought upon ourselves here on Earth: violent political conflict, pronounced economic disparity, destruction of the planetary environment, etc.

A cooperative approach towards future Space exploration would entail very different strategies from ones that are competition-driven. Nations would collaborate, sharing rather than hoarding technical expertise so that each one would benefit from the contributions of others, and all of them together would benefit by reaping the greater collective rewards that are not possible through isolated, unilateral efforts. Space settlers would not necessarily attempt to replicate an Earth-like ecosystem on Mars, but would support whatever indigenous life may be found there, despite its genesis. The same consideration would be extended to the inorganic landscape as well, whether life there exists or not. Humans would live on Mars in a way deemed to be harmonious with the planet, and not what it 'should' be in order to support humans. To live harmoniously in Space means to understand

E.S. Lockard, *Human Migration to Space*, Springer Theses,
DOI 10.1007/978-3-319-05930-3_5,

how our extended presence positively affects—rather than 'contaminates'—the extraterrestrial venues we desire to occupy. Cooperative strategies can be implemented in two realms: the use of technology to interface with the Space environment; and the establishment of seminal social structures.

The justification for cooperation-based models can be found in various related disciplines: Margulis, Kropotkin and Dawkins have written extensively about cooperation in evolution theory; Mander, Hawley, Steiner, McKenzie, Sale, Devall and Sessions, discuss cooperation in ecology theory; and Oye, Taylor, Cronin, Steiner, Maruyama, and Axelrod write about cooperation from a socio-political perspective.

5.2 Theories of Evolution

5.2.1 Traditional Doctrine: Natural Selection

Any discussion of evolutionary theory inevitably references Darwin, whose theory of natural selection still reigns as the dominant paradigm in the field of biology. So dominant is the model that the principles of natural selection have pervaded fields far beyond the life sciences, providing the underlying justifications for much of modern social, economic, and political theory.[1] In fact, hardly any aspect of Western culture has not been shaped by the notion of 'survival of the fittest' in some way, although the concept is often misappropriated from its biological roots. When it was introduced, Darwin's theory of the derivation of species revolutionized the way in which the natural world was understood, by revealing the connections between all living things.[2]

Though *The Origin of Species* was Darwin's seminal work, his subsequent tome, *The Descent of Man*, delved specifically into human evolution—a subject which was consciously omitted in his magnum opus.[3] While it was this work that laid out the premises for our origins in the primate world as well as for sexual selection as the basis for racial divergence, his primary motivation for writing the book—arising from the political and social milieu of the times—was to argue from a scientific perspective for the equality of the races.[4] This historical context is critical in

[1] A few such examples include: Ayn Rand's ethical egoism, the American bipartisan system of democracy, capitalism and the free market, and the eugenics movement.

[2] In addition to which Neo-Darwinism revealed through genetic theory the *similarities* between all living things.

[3] Darwin rightfully feared that his first volume, in contradicting the creationist beliefs of the Church, would be perceived as heretical and that there would be a significant backlash against his writings; thus he chose not to include humans in the earlier formulations of his theory.

[4] Darwin was a resolute abolitionist, and his political agenda in his scientific endeavors stemmed from his abhorrence of slavery.

interpreting Darwin's writings because the thrust of his work was as political as it was scientific.

Heavily influenced by Malthus' writings on the relationship between population growth and subsistence, as well as Hobbes' views of the state of human nature, Darwin saw the world as an arena of individual struggle and unforgiving brutality, thus framing his speculations on the emergence of superiority through the process of ruthless competition. Survival of the fittest implied that those who were best equipped to adapt to their harsh and merciless environs were more likely to survive and reproduce than those who were 'weak' thus the advantageous traits possessed by the 'stronger' individuals would be passed on to its offspring.

The inference reflects the patriarchal thinking of that time: whereas masculine character traits such as ambition, physical strength, and individualistic egoism were held to be of higher value because they bestowed a competitive advantage in his model of progress, the typical feminine characteristics of mutual support, nurturing, and empathy were not considered favorable to competition; hence women in Darwin's view were inferior because they did not possess the necessary attributes that give rise to evolutionary progress as he defined it (Darwin 2004, p. xlviii).[5] In brushing aside 'feminine' traits, Darwin's theory has perpetuated patriarchal worldviews that are becoming untenable. A new way of thinking about evolution must also identify strengths in typically feminine attributes.

In the wake of Crick and Watson's discovery of DNA, neo-Darwinists built upon the fundamental tenets of natural selection, positing the gene as the unit of variation that is passed on to the next generation, not entire traits as Darwin had thought. If random genetic mutations conferred an environmental advantage to the organism, then the organism possessing it would be more likely to survive and reproduce, thus passing that mutation on to its offspring. However, random mutations in a few individuals alone are not sufficient to constitute genetic change; there must be a critical mass of a population—an indication that the inherited changes are resilient and enduring—who eventually possess the same variations in order for it to constitute evolutionary change.

In demonstrating both the derivation and the interrelatedness of species, the way in which the natural world was understood was transformed. Darwin overturned the prevailing worldview through the science of biology—as did Copernicus in the field of astronomy and Newton in the field of physics before him—and like most intellectual upheavals in human history, clashed with theological doctrine of that time. While Copernicus' discovery threatened our place in the Universe, and Newton's natural law challenged the laws of God, Darwin's theory—in revealing human's primate ancestry—undermined the belief in our uniqueness, and dethroned humankind as the master race on the planet. Ours was no longer the privileged species made in the likeness of an omnipotent higher being, but instead

[5] Patriarchal attitudes are prevalent throughout Darwin's reasoning, and are a marked contrast to subsequent developments in evolutionary theory which reflect a more feminized view of survival, novelty, and robustness.

descended from the inferior caste of lower beasts. Though Darwinism did nothing to deflate the myth of human superiority, the theory of life as a gradual continuum of change made it clear that we were also by no means the final or ultimate species.

Darwinism is limited however, in that it never adequately explains a wider range of biological phenomena, such as the process of speciation; it offers only gradual accumulated change which at best gives rise to variation. Assuming that the incremental changes resulting from individual mutations are cumulative, even radical developments such as the eye are difficult to account for. In the fossil record—the evidence of life's history—evolutionary discontinuities are rife. While proponents of Darwinism argue that these discrepancies are due to the fact that the fossil record is incomplete—and therefore inconclusive—other scientists, such as Stephen Jay Gould and Niles Eldredge, have proposed another explanation to address this deficiency. Their theory of punctuated equilibrium posits that pronounced morphological changes are advanced in leaps—not incrementally—although it is not clear as to the causes of these leaps other than genetic isolation. Another theory of evolution picks up where natural selection left off.

5.2.2 Alternative Paradigm: Symbiogenesis

Mounting evidence over the last 40 years suggests that there are more compelling forces at work in generating genetic change, and paint a radically different picture of the complexification of life than Darwin's natural selection. In her 1966 thesis, biologist Lynn Margulis posited symbiosis as the *primary* force in evolution. Although her theory on symbiogenesis was originally dismissed by the scientific community, it has increasingly found favor in the wake of our deepening understanding of biological forces, validated by evidence in the Human Genome Project (Thieman and Palladino 2007)—which confirmed that our genetic origins can be traced back to single-celled organisms—as well as in the discovery of lateral gene transfer.[6] The theory of symbiogenesis is more persuasive than Darwinism in accounting for complexity and novelty in evolutionary processes, as well as for cooperation and reciprocity that is manifest in the behavior of all living (and even non-living) things.[7]

In natural selection, genetic change is accrued through random mutations: the mutations which better respond to the challenges of the environment confer a reproductive advantage to an organism, who is then more likely to survive and pass that mutation onto its offspring. As more of the organisms who possess an advantageous mutation survive and subsequently reproduce, the more populous they become, eventually replacing those who have less advantageous traits.

[6] www.genomenewsnetwork.org

[7] Darwin observed cooperative behaviors both between and within species that could not be adequately explained by his theory of natural selection.

In contrast to natural selection, symbiogenesis is the process by which complex organs, new organisms, and even new species come into existence through the symbiotic merger of two independent, heterogeneous entities. Supplanting the vertical transfer of genes from one organism to its offspring through sexual reproduction, two individuals can directly exchange and combine genetic material, with the resulting organism possessing the entire DNA structures of both its predecessors. Margulis explains that "through symbiosis, organisms acquire not traits but entire other organisms, and their entire set of genes…These become new individuals at larger, more inclusive levels of integration" (Margulis 1998, p. 8). Symbiogenesis, explains Margulis, allows for much greater complexity and integration of systems through bilateral genetic exchange (known as recombination) than the linear, incremental process of accumulated change through single chance mutations. According to physicist Stephen Wolfram, "the Darwinian theory of evolution by natural selection is often assumed to explain the complexity we see in biological systems…but it has never been at all clear just why this theory should imply that complexity is generated. [In fact] in many respects it tends to oppose complexity" (Gardner 2003, p. 201).

Margulis' theory also offers a more convincing explanation for the abrupt emergence of evolutionary novelty than classic Darwinian theory does. The fossil record indicates that the majority of species remain in a state of evolutionary stasis for most of their existence, followed by radical transformation (i.e., variation or speciation) in a relatively short period of time. The gradual accumulative change that is characteristic of natural selection is not sufficient to explain this phenomenon. But because recombination is a much faster and more extensive process of change than isolated mutations, "symbiosis as a source of evolutionary novelty helps explain the observation of 'punctuated equilibrium,' of discontinuities in the fossil record" (Margulis 1998, p. 8).

The concept of co-evolution through symbiotic interaction is also very different from that of natural selection, in that it is inherently an additive process rather than a subtractive one. In the co-evolution of predator and prey from a Darwinian perspective, any imbalance in one population eventually causes a subsequent inverse imbalance in the other: if, for example, the prey acquires traits that enable it to better evade its pursuer, then either their population will increase—producing more subjects for the decreasing numbers of predators—or it will trigger adjustments among the predator population to become more agile in pursuit. In either case, the number of predators increase, and the process occurs again in the reverse (Dawkins 2006). In symbiotic theory, co-evolution is understood as a mutual exchange of traits between organisms (or species) that are *not* in competition with one another. Change does not occur through the *chance mutation* of traits but through the *reciprocal sharing* of traits. The notion of fitness is not a relative or selective condition as it is in Darwin's theory; bonds are forged through diversity of traits between organisms with cooperative dispositions. Though a "relationship of symbiosis may begin as a parasitic connection, once the interactive process of development is established, co-evolutionary possibilities are opened up that may lead to radically new adaptive capabilities" (Goodall 2005, p. 29).

This updated view of evolution through symbiotic transformation stands in stark contrast to classical evolutionary theory: whereas natural selection is based—in a brutal and dispassionate manner—on zero-sum principles of competition in which the so-called strongest/fittest members within a population are those most likely to reproduce (thus leaving the weaker or less fit to perish),[8] symbiosis is based on non-zero sum principles of cooperation between members within a population as well as between species, acknowledging and rewarding diversity. In symbiotic relationships, there are no individual 'winners' and 'losers' as is the case in natural selection; each entity potentially has something to offer which benefits another. Fitness, therefore, is not based on a dichotomy of strength and weakness, but rather on the recognition of a critical diversity between entities at various scales that establishes a basis for the exchange of advantageous traits. This reinterpretation of fitness is also held by Kropotkin, who argues that "those animals which acquire habits of mutual aid are undoubtedly the fittest" (Kropotkin 2006, p. 5). The view of evolution as "chronic bloody competition. . .dissolves before a new model of continual cooperation, strong interaction, and mutual dependence among life forms." Margulis contends that "life did not take over the globe by combat [as is depicted in natural selection], but by networking. Life forms multiplied and complexified by co-opting others" (Margulis and Sagan 1997a, p. 29).

Although symbiotic theories of evolution diverge fundamentally from Darwinian theory, the concept of symbiogenesis is not irreconcilable with natural selection. Rather, the two types of processes are applicable at different scales of organismic interaction; both competition and cooperation are evolutionary forces, and thus should be seen as complementary—not antagonistic—to one another. What remains in dispute however, is which process is the primary driver of evolutionary change, and which has a more subordinate role in the process. Margulis boldly flips the traditional paradigm on its head: "symbiogenesis is the author [of evolutionary change], natural selection is the editor" (Kelly 1994, p. 373). "Mutual aid is as much a law of animal life as mutual struggle, but. . .as a factor of evolution, it most probably has a far greater importance, inasmuch as it favors the development of such habits and characters as insure the maintenance and further development of the species" (Kropotkin 2006, p. 5). Dyson echoes the same position: "What leads organisms to evolve to higher types? Darwinian evolution, as Stephen J. Gould among others has pointed out, does not 'progress' toward greater complexity, but Darwinian evolution plus symbiogenesis does. . .Darwinian evolution, in one of those paradoxes with which life abounds, may be a victim of its own success, unable to keep up with the non-Darwinian processes that it has spawned" (Ward 2001, p. 149). The continual increase of interdependence between living things—currently over 50% of all species are parasitic (Kelly 1994, p. 75)—indicates that

[8] Margulis claims that popular interpretation of 'survival of the fittest' is a misunderstanding of the theory; fittest does not necessarily refer to the physically strongest, but simply those most likely to reproduce—which therefore makes it a tautological statement (1998).

evolutionary processes themselves are evolving towards greater interdependence and cooperation.

The differences between the theories of symbiosis and Darwinism are analogous to those between quantum and Newtonian physics, the discovery of which also ushered in a revolutionary shift in worldview. While the Newtonian paradigm is still relevant in explaining the behavior of macro-phenomena, at the micro level the deterministic principles of cause and effect no longer apply. Quantum mechanics has revealed that subatomic behavior is inherently chaotic as opposed to ordered, unpredictable instead of deterministic, and based on probabilities rather than certainties. This revelation has had profound implications for our worldview; we no longer live in a Universe that is rational, quantifiable, and dependable, but one that is random, paradoxical, and even alterable through mere observation. If quantum theory, in challenging both our epistemological and ontological foundations, represents a 'physics' revolution, then symbiotic theory, in challenging the way in which we understand how life has evolved, represents an 'evolution' revolution. Freud once stated that "all great scientific revolutions feature two components: an intellectual reformulation of physical reality and a visceral demotion of *Homo sapiens* from arrogant domination" (Gardner 2003, p. 127). Where Darwinian theory once demoted *Homo sapiens* as the unique and superior species on the planet, symbiogenesis has relegated natural selection to a subordinate role in the processes of evolution by demonstrating that life has been created and perpetuated primarily through alliances—not through adversarialism.

The principles behind symbiotic theory are appealing in that they entail much greater potential not just for *transformation* but *transcendence* of life as well, and may very well reshape our worldview once again. Evidence of this can already be found in contemporary theories of technology as well as the literature on cyborgism and the posthuman, all of which take cues from symbiotic doctrine. With each reconfigured worldview, the concept of what it means to be human is irrevocably altered, as is our relationship to the Universe as well.

5.2.3 A Comparison of Evolutionary Worldviews

The Darwinian sensibility in the current western socio-political ethos places value on individual gain at the expense of others, on hierarchical positions of dominance and authority, and on conquest over perceived adversaries. Natural selection results in the ascendance of some based on the descent of the majority; success and fulfillment are therefore only for a privileged few who are then considered superior to those that have been defeated. There is no justification for acts of compassion or altruism, as altruistic behavior is seen as detrimental to the interests of the individual. In the Darwinian world, one can only succeed at the expense of others.

In contrast, the symbiotic sensibility espouses mutual benefits for all individuals, coalition-building through lateral and interdependent networks, and cooperation between heterogeneous entities. Symbiosis relies on (as well as generates) diversity within and between populations, creating scales of kinship and hybridization which in turn leads to greater complexity. In fact, "better conditions are created by the elimination of competition by means of mutual aid and mutual support" (Kropotkin 2006, p. 60). Whereas "natural selection primarily means the destruction of the unfit" (Kelly 1994, p. 370), the tendency of which results in genetic uniformity, "symbiosis on a large scale could drive many of the complexities in nature that seem to require multiple simultaneous innovations. . .it would exploit the power of cooperation rather than competition, exclusively. . .cooperation nurtures a type of diversity that competition cannot produce" (Kelly 1994, p. 372).

If diversification of life can not only better withstand ecological crises and other catastrophes, but also provide the catalyst for the transformation of life, then both diversity and the processes of diversification can inform more robust and sustainable relationships outside the biological sphere as well. "The recognition of symbiosis as a major evolutionary force has profound philosophical implications. All larger organisms. . .are living testimonies to the fact that destructive practices do not work in the long run. In the end the aggressors always destroy themselves, making way for others who know how to cooperate and get along. Life is much less a competitive struggle for survival than a triumph of cooperation and creativity" (Capra 1996, p. 243; Kropotkin 2006, p. 242).

While neo-Darwinists argue that altruism and commensalism (which entail acts of good will and sacrifice) are inconsistent with the gene's disposition towards selfishness, reciprocal action (which requires no selfless intent or sacrifice) may in fact, better serve the gene than competition does. Though mutualism is critical to the symbiotic function, it is not dependent on compassionate motivation; it can stem from entirely selfish drives. Natural selection pits one against another, thereby 'weeding out' organisms that are not competitively successful, both in terms of individual survival and perpetuation of a gene line. Symbiosis—in exchanging genes rather than eliminating them—reinforces collaboration between individual organisms (insuring immediate survival) and species (insuring long-term, collective survival), allowing for far more complex relationships and emergent properties to develop than competitive practices which are essentially reductivistic. Competition is nihilistic towards relationships by thinning out the population; in contrast, symbiosis is a relationship-building process. Progress and achievement are not defined by the outstanding accomplishments of a single individual but by the aggregate of the entire community. Mutual aid is "the surest means for giving to each and to all the greatest safety, the best guarantee of existence and progress, bodily, intellectual, and moral" (Kropotkin 2006, p. 61).

The contrasting theories of evolution paint a different picture of the connectedness of life. Darwinian natural selection demonstrates the connection between living things through hierarchicalized linear progression; genetic theory reveals the similarities and commonalities between living things at the molecular and cellular level; and symbiotic theory uncovers the interdependence between disparate forms of life, as well as the autopoietic nature of those relationships, in which complexity develops through lateral and networked interactions.

The symbiotic perspective also radically undermines the Cartesian/Newtonian conception of the human subject as an objective and independent entity that is distinct from Nature. Though the Darwinian revolution dispelled the myth of our uniqueness, it nevertheless solidified our belief in our superiority over all other 'lower' life forms. In turn, "we have extended this fragmented view to our human society, dividing it into different nations, races, religious and political groups. The belief that all these fragments—in ourselves, in our environment, and in our society—are really separate has alienated us from nature and from our fellow human beings and thus has diminished us" (Capra 1996, p. 296).

Cooperation, as symbiotic theory demonstrates, presents possible strategies for the development of our relationships to our technologies as well as to our environments in Space that promise maximum collective benefit. Through its reliance on diversity, alliance-building enhances the outlook for long-term survival by promoting robustness, resilience, and endurance.

5.3 Theories of Environment

Understanding humans as an integral part of their environment—as much so as amoeba or rocks—is key to my argument about adaptation in Space. Our view of ourselves as separate from Nature/Space, and the use of technology as antithetical to Nature, is the root of our alienation and our sense of hostility towards it. Technology can be implemented to create greater empathy and kinship through intensifying the experience of the environment, not controlling it or shielding us from it. This allows for the implementation of some technologies however, that we may not feel altogether comfortable with—like terraforming—but it is absurd to think we can inhabit new 'ecological' niches without our presence having some impact on them. The question is not whether or not we should do it, but how we can go about it in a way that is advantageous to all. Instead of thinking of ourselves as 'contaminating' the Space environment (which is what we have, in fact, done through our current approach in which we produce and dispose of toxic wastes) we should consider how our presence can be beneficial—not to just to us, but to the environment as well—and then develop strategies to achieve that.

How humans perceive themselves and their relationship to the natural environment is still predominantly influenced by the deeply entrenched theory of natural selection, but symbiotic theory is increasingly challenging the Darwinian worldview, offering a cooperative perspective as an alternative to one of competition and conquest.

Although Darwin's theory made clear that we are not the final species in the evolutionary chain, many humans still cling to the belief that we represent the pinnacle of evolution on Earth. Change has been traditionally understood as a linear and progressive process, where the transition from one species to the next implies hierarchical status between organisms—specifically, the superiority of successor over predecessor. In this view—which emerged during the Enlightenment—*Homo sapiens* are elevated above all other biota, subjugating Nature as their domain to rule, conquer, or nurture as they see fit. Humans are not an integral part of Nature but reside in a sphere separate from and above it; no longer is our role one of participation in its processes but instead one of manipulation and control.

Symbiotic theory posits the relationship between humans and environment in a qualitatively different manner from the Darwinian interpretation. In the symbiotic paradigm, *Homo sapiens* are understood as an *extension* rather than as a *progression* of microbial evolution. The human is not a more complex organism than a bacterium, just a more complex *recombination* of microbial organisms; hence, it is erroneous to define ourselves as distinct from them or any other species—our bodies alone are comprised of more bacteria than genetic human cells (Margulis). Without the multitudes of bacteria that inhabit our bodies, we could not survive; bacteria, on the other hand, would certainly continue to thrive without us. We are literally comprised of organisms that are ubiquitous in our environment, and thus it might be more accurate to describe all eukaryotic (nucleated cell) life as a macrocosm of microbes, and the planet as a "superorganism of bacteria" (Margulis and Sagan 1997b, p. 31). The notion of evolution as progression gives humans license to claim intelligence and technology as their own, whereas evolution as extension would describe attributes of a species as shared with all other species. Moreover, the origins of those attributes do not reside within the species in which they first appeared, but in the very first forms of life. They belong to the larger biotic collective and not to a single species. "Intelligence and technology, incubated by humankind, are really the property of the microcosm" (Margulis and Sagan 1997b, p. 36).

It is the single-celled organisms that have dominated the history of life on Earth, and are the primary force that maintains ecological balance within the biosphere. However, Margulis rejects the temptation to simply reverse the Darwinian hierarchy and reposition humans as inferior to other forms of life (as Derrida suggested in his theory of deconstruction), arguing that this shift would only perpetuate the stratified, dualistic, and antagonistic character of the model that she seeks to reconceptualize (Margulis and Sagan 1997b, p. 18). Humans are as integral to Nature as all other forms of life. Understanding evolution as a process of lateral extension rather than of hierarchical progression finds parallels in other theories of environment as well.

5.3.1 Deep Ecology

The symbiotic view is ingrained in the deep ecology philosophy, a basic tenet of which is that "humans are 'plain citizens' of the biotic community, not lord and master over all other species" (Devall and Sessions 1985, p. 68). This is a marked departure from the traditional depiction of humankind, in that it subsumes the human species under Nature, rather than elevating us above it. Capra writes that "deep ecology does not separate humans from the natural environment. It sees the world not as a collection of isolated objects, but as a network of phenomena that are fundamentally interconnected and interdependent...[it] recognizes the intrinsic value of all living things" (Capra 1996, pp. 6–7). Biocentric equality, which acknowledges that all living things in the biosphere have intrinsic value as well as an equal right to 'self-realization,' is a basic tenet of deep ecology (Devall and Sessions 1985). Like symbiotic theory, there is no hierarchical stratification of living things.

5.3.2 Gaia Theory

James Lovelock's Gaia theory is built on the same premise as symbiotic theory, but extending the principles beyond the biological realm to the geological scale. Living things do not simply react to the vicissitudes of their surroundings, but actively alter them as well, so that both animate life and its inanimate environment, both the organic and the inorganic, are constantly responding and adapting to each other in an interactive feedback loop. Life and its environment grow increasingly complex synchronously in an autopoietic process: as life evolves it participates in the evolution of its environment, just as ecological changes precipitate the evolution of its biota. Life and the environment are constantly adapting to each other, from scales that are imperceptibly small to scales that are incomprehensively large. Lovelock observed that "so closely coupled is the evolution of living organisms with the evolution of their environment that together they constitute a single evolutionary process" (Capra 1996, p. 227). Therefore "an individual and its environment [are] best understood not as separate entities but as a dynamic feedback system" (Gallagher 2001, p. 103). According to Lovelock's hypothesis, the presence of life modulates ecological systems, which in turn maintains conditions that support and modulate life. The Earth is not fixed and static as we may observe at any given moment, but rather in a state of homeostasis—a dynamic, but stable, disequilibrium (Lovelock 2000). It is best characterized not as a condition of stability, but one of stabilization; in other words, a *process* of becoming rather than a *state* of being. Lovelock concludes that this condition of disequilibrium is evidence for the existence of life and conversely, equilibrium is synonymous with its absence.[9]

[9] Lovelock developed this position when asked by NASA to determine if there were signs of possible life on Mars; because he observed no volatility in its atmospheric gases, he concluded that the planet, being in a state of equilibrium, offered no evidence for the existence of life there.

5.3.3 Cosmic Evolution

In the future we may be able to discover similar Gaian relationships on a galactic, or even cosmological scale, a view that is held by Fritjof Capra, who regards the Earth—with all its life on it—and the rest of the Universe, as a coherent, mutualistic entity; "outer space and Earth's interior are both part of Gaia's environment" (Capra 1996, p. 214). As we venture further into Space and learn more about our celestial surroundings, it stands to reason that we are likely to find increasing evidence of macroscale reflexive processes. We see evidence already on the scale of the solar system; planetary conditions are affected by their interactions with other neighboring masses as well as by the Sun. Tidal cycles on Earth are influenced by the Moon's gravitational forces. Orbits change according to proximity to other nearby planets. It is believed that liquid water oceans exist under the water-ice crust of Europa due to frictional stresses caused by its proximity to Jupiter and her other satellites, Io and Callisto. Once we learn more about cosmic conditions beyond our solar system, we are likely to find similar relationships. These discoveries make it clear that no planetary body is exempt from the influence of other entities in its vicinity. Change occurs not just at the biological or geographical scale, but also the cosmological scale. Without the presence of dynamic forces that upset the state of equilibrium, "cosmic evolution itself would cease" (Chaisson 2001, p. 29).

5.3.4 Selfish Biocosm Hypothesis (Strong Anthropic Principle)

The selfish biocosm hypothesis not only also posits that the entire Universe is evolving, but that the Universe is *living*. Gardner's hypothesis is based on the strong anthropic principle, which holds that evolutionary processes did not incidentally give rise to life, but actually work in favor of life. Echoing similar themes in Gaia and symbiotic theory, he argues that "so fined-tuned are the laws of physics to favor life that it almost seems as if the universe and the life forms that inhabit it have coevolved, like earthly creatures and the ecosystems they populate" (Gardner 2003, p. 15).[10]

Though his hypothesis seems to be at odds with the principles of natural selection, "this attempted reconception of the nature of Nature is emphatically not an attempt to overthrow Darwinism, any more than is the Stephen Jay Gould/ Niles Eldredge vision of punctuated equilibrium, Lynn Margulis' notion of species cooperation as an evolutionary strategy, or Stuart Kauffman's idea of

[10] Though the scientific community has an expressed aversion to the teleological undertones of such a position, our planet has endured many cataclysmic events in which all existing life on its surface was nearly wiped out, and yet new types of organisms emerged to populate the Earth once again.

self-organization as an engine of evolutionary advancement. On the contrary, it is an effort to take seriously the radical possibility that evolution. . .operates on an unexpectedly grand scale" (Gardner 2003, p. 95).

Gardner's thesis also draws from Margulis' theory: "The evidence that symbiogenesis has profoundly shaped evolutionary history is, in fact, pervasive at all levels of emergence—from the appearance of eukaryotic cells to. . .the emergence of advanced mental functions is an artifact of human social interaction. . .the symbiogenesis approach. . .may hold the key to understanding the dynamics that will propel the next great evolutionary leap forward, which is likely to be fueled by a technology-mediated (and thus vastly accelerated) version of Wilson's gene/culture evolution" (Gardner 2003, pp. 196–197). Gardner concludes that "broad-based altruism is, in effect, a cosmic necessity" (Gardner 2003, p. 224).

The various theories of evolution and the environment seem to support the notion that cooperation is prevalent not only on Earth, but throughout the Universe.

5.4 Advantages of Cooperation

There are three basic types of interaction between species according to Steiner's fifth principle of ecology: competition, symbiosis, and predation. In predation/parasitism "the outcome benefits one and is detrimental to the other. Competition leads to negative outcomes to both groups involved. . .[but] symbiosis [is the only one of the three types of interaction] which benefits both participants" (Steiner 2002, p. 31). A comparison of behaviors ranging from competitive to cooperative illustrates the outcomes associated with the various types of interactions:

Detrimental:	Competition = both parties are harmed (– –)
	Amensalism = disadvantageous to one party while the other is not affected (– 0)
Break-even:	Neutralism = both parties are unaffected (0 0)
	Parasitism = one party benefits while the other is harmed (+ –)
Beneficial:	Commensalism = one party benefits while the other suffers no disadvantage (+ 0)
	Mutualism = two different parties benefit and neither suffers (++)

Of all the options, reciprocity is the only type of interaction that results in positive outcomes for all participants. Not only does mutualism maximize the collective good, it is also the most advantageous strategy from an evolutionary perspective. In his book, *The Evolution of Cooperation*, Robert Axelrod presents several strong arguments that prove cooperation is an evolutionary stable strategy (ESS), and why cooperation is preferable over competition as the way to get ahead in evolution. He also exposes false assumptions about the conditions necessary for cooperation to take hold, by demonstrating that "cooperation can get started and prove stable in situations which otherwise appear extraordinarily unpromising. . .-friendship is hardly necessary for the development of cooperation" (Axelrod 1987, pp. 21–22). Axelrod redefines the notion of biological fitness in non-Darwinian terms: "an individual able to achieve a beneficial response from another is more likely to have offspring that survive and that continue the pattern of behavior which elicited beneficial responses from others" (Axelrod 1987, pp. 21–22).

Cooperative behavior flourishes best in relations of kinship where compassion and empathy are present: however, an altruistic disposition is not a prerequisite condition; Axelrod observes that although "almost all clear cases of altruism, and most observed cooperation—apart from their appearance in the human species—occur in contexts of high relatedness, usually between immediate family members. . .conspicuous examples of cooperation also occur where relatedness is low or absent" (Axelrod 1987, pp. 21–22). Even if the gene is indeed entirely 'selfish' and acts only in its best interests (Dawkins 2006), cooperation can still prove to be the most viable strategy. Contrary to common assumptions, "the foundation of cooperation is not really trust, but the durability of the relationship" (Taylor 1976, p. 182). What is meant by durability has nothing to do with the strength of bonds between individuals that Axelrod alludes to, but rather the *continued prospect* of engagement with one another (Taylor 1976). There must be 'a sufficiently long shadow of the future,' as members are more inclined to cooperate if they believe they will encounter each other again in the foreseeable future (Axelrod 1987). In large populations where anonymous exchanges are the norm, cooperative behavior is less likely to be fostered because members cannot identify each other and the chance of repeated encounters is also much smaller. This suggests that communal populations should be small enough in number so that everyone can be acquainted with each other. If the population is too dense such that there is the reduced likelihood of subsequent encounters, the shadow of the future is diminished, resulting in a stronger temptation to act only in accordance with one's immediate self-interest. But however, if the shadow is too long, the incentives for immediate gain will more likely outweigh the greater incentives for long-term or distant gains. Thus if the right conditions exist, political will or authoritative oversight is not necessary for enforcing cooperative behavior; in fact, the presence of a central authority actually tends to undermine the conditions that support cooperation, because governments effectively dictate such behavior through policy, rather than encouraging it through negotiation (Taylor 1976).

The most effective strategy to elicit cooperation is Tit-for-Tat (Dawkins 2006; Axelrod 1987). It is a surprisingly simple strategy as well, based on values of friendliness and forgiveness. The principles are as follows:

1. The first move should always be to cooperate. (friendliness)
2. If a player defects, the next move should also be to defect. (retaliation)
3. If the player's next move returns to cooperation, resume cooperation as well. (forgiveness)

"Tit-for-tat's robust success is due to being nice, provokable (retaliatory), forgiving, and clear. Its retaliation discourages the other side from persisting whenever defection is tried. Its forgiveness helps restore mutual cooperation. And its clarity makes its behavioral pattern easy to recognize" (Taylor 1976, p. 176). It is both an elegant and appealing strategy: "Tit-for-tat succeeds without doing better

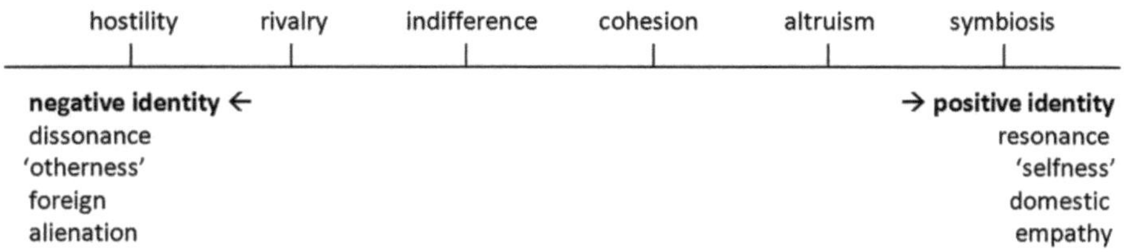

Fig. 5.1 Continuum of identity

than anyone with whom it interacts. It succeeds by eliciting cooperation from others, not by defeating them. [It advises:] don't be envious, don't be the first to defect, reciprocate both cooperation and defection, and don't be too clever" (Taylor 1976, p. 190). A strategy of 'Always defect' is also an evolutionary stable strategy, so in order for Tit-for-Tat to become the dominant strategy in a population, it is important to create population clusters that are conducive to cooperation amongst themselves; that is, to establish pockets of kinship.

Although cooperation does not require relations of kinship to be present as a necessary condition, cooperative behaviors can actually establish new affiliations through shifting perception of identity relations. Identity is not fixed and discrete but rather fluid and analogous; distinctions between 'self' and 'other' are actually gradations along a dynamic continuum. Taylor ascribes types of social relations to various levels of identity (Fig. 5.1). He describes identity as:

> . . .a continuum from negative to positive, ranging from conceiving the other as the social or cultural opposite of the self to viewing the other as an extension of self. Building from this premise, a pure-positive social identity can be defined as *symbiosis*, a relationship in which the actors view each other as extensions of themselves. A highly positive but less intense identification can be defined as *altruism*, where the actors retain their individuality but are so closely identified with another that they are willing to make sacrifices on their behalf. The next level on the continuum is *cohesion*, a situation in which actors recognize the common good among themselves and view themselves as part of a conceptual group. The middle level is *indifference*; the actors are not important enough to each other to have any positive or negative evaluations. A moderate form of negativity is *rivalry*; others are seen as competitors and objects for the fulfillment of self-interest. The most intense negative identification is *hostility*, a situation where the actors hold a mutual antagonism to the point of seeing each other as the 'anti-self' or enemy (Taylor 1982, p. 17).

This continuum of identity, in reflecting degrees of kinship or affinity, has bearing on propensities for cooperative behavior. "There is strong correlation between the form of reciprocity and kinship distance: as the kinship distance separating two individuals increases, the more do relations between them incline towards the unsociable negative end of the spectrum. Kinship groups. . .can be thought of as a series of nested sectors. Generalized reciprocity is confined to the household and local lineage sectors, balanced reciprocity mainly to the village and tribal sectors, and negative reciprocity to the intertribal sector. . .the negative reciprocity usually characteristic of intertribal relations may be replaced by a more balanced reciprocity in cases where there is trading or other intertribal symbiosis" (Taylor 1982, p. 67).[11]

Since the tendency towards reciprocity diminishes with perceptions of social distance, another way by which to reinforce the disposition towards cooperation would be to expand or recontextualize scales of kinship. The Overview Effect is an example of how perceptions of affiliation can change through scalar expansion. The astronauts who viewed the Earth from afar suddenly saw themselves as part of a larger unity, one that transcended race or nationality, and reinforced their bonds with all of humanity. Situated in the overwhelming vastness and emptiness of Space, apart from all other human beings—from all life for that matter—they experienced a deeper connection to the entirety of the Earth, with all things living. This phenomenon demonstrates that scales of kinship can broaden as long as there is the perception of a phenomenologically external entity that constitutes 'other,' one that reinscribes a more inclusive circle of empathy and thus the boundaries of identity. Through expanding scales of affinity, "the individual feels a sense of belonging, of connectedness, to the cosmos as a whole" (Capra 1996, p. 7).

Displacement—a phenomenological uprooting from the conditions of the familiar (where one 'belongs'), and immersion into conditions of otherness (where one doesn't 'belong')—can also be seen as a catalyst for internal bonding in the isolated and confined environments of Space, which provide fertile conditions for cooperation to arise: "the coherence of the collectivity is strengthened by a threatening environment, and two things above all tend to make an environment threatening: first, the territory of the society in question is circumscribed by such natural features as mountains, deserts and seas, so that expansion and flight are impossible; second, there are enemy or rival societies nearby" (Taylor 1982, p. 135). Whether on Mars, on the Moon, or in LEO, there is no option to flee from dangerous circumstances, and while there are no 'enemies' in the vicinity, external conditions pose a constant threat to the safety of the crew. However, depicting the Martian environment as adversarial is counterproductive, since that will only exacerbate the feelings of alienation and undermine the process of adaptation to it.

Cooperative approaches require that all entities in the environment are acknowledged for their intrinsic value—that each has something to give to and to gain from the other, either directly or indirectly. Although it may seem odd to ascribe agency to simpler life forms like amoeba or micro-organisms, and even more so to inanimate entities like rocks, both symbiotic theory and Gaia theory have demonstrated that cooperative, autopoietic relationships at these levels indeed already exist. Cooperation with the environment includes not just living entities but the non-living as well. "Life likely differs from the rest of clumped matter only in degree, not in kind… [there is] no special life force that would set animate beings manifestly apart from all other forms of inanimate complexity" (Chaisson 2001, p. 122).

[11] This scalar expansion of kinship is what Lanier refers to in his 'circle of empathy' which inscribes the domain of those we consider deserving of ethical treatment. The greater the sense of identification we feel towards something, the more we feel empathy for it (Lanier 2010).

The conditions then that would need to be established in Space in order to promote cooperation both socially and environmentally are the following:

- Generation of empathy through expanded scales of kinship
- Sufficiently long shadow of the future for interactions
- Durability of the relationship
- Regular and non-anonymous encounters (accountability)
- Appropriate size of field of interaction
- Integrative approach that fosters engagement and negotiation (opportunity)

5.5 Prescriptions for the Development of Cooperative Relationships in Space

5.5.1 Cooperative Strategies for Technological Adaptation

While the use of any particular technology is dependent in part on its inherent capacities, it is also subject in part to social constructions in which human intentionality plays a significant role in how they are implemented. The prospective technologies that are applicable to living in Space may mediate in ways that are exploitative and antagonistic; or they may mediate to encourage cooperative and integrative behaviors. It is not sufficient to simply recommend a type of technology to be developed, but to also prescribe *how* they should be developed. It must be clear first what purpose they are intended to serve.

5.5.1.1 Cyborg Technologies

The cyborg route entails less risk and avails us of many more options once we inhabit Space. It is arguably the best alternative in terms of technologies that can create a phenomenological sense of 'belonging to a place' and that can amplify the experience of the environment.

It also poses less demands on the habitat and other 'survival' technologies; in fact, if humans are engineered to be able to tolerate the conditions of the extraterrestrial environment, then there will no longer be the need to create shields or protective barriers that are intended to segregate humans from their environs. Conditions on Mars or the Moon would no longer be characterized as 'harmful,' 'dangerous,' or 'life-threatening.' Once freed from these obligations, the applications of Space technologies can be elevated to higher-order purposes, like satisfying aesthetic urges, participation in the arts, achieving self-fulfillment, and developing altruistic behaviors.

Cyborg technologies should be implemented not with the intent to subordinate humans to the environment or to treat the extraterrestrial environment as one that has to be *deferred to* (that is, putting the interests of the environment over those of humans, assuming that they are even exclusive to one another), but as something to be *engaged with*. Cyber technologies should be employed as a means not only to survive but to better experience the physical condition of their new world, and to be an integral part of Nature in Space. Application of integrative cyborg technologies are those that make us more mobile, reduce the harmful effects of micro-gravity and radiation on the human body, enhance sensory perception, and allow greater exposure to minimal atmospheric pressure, colder temperatures, and different air compositions. "Future Moon or Mars walkers—or orbiters—can be expected to have much better prosthetics that will enable them to see, hear, feel, perhaps taste and smell, and to walk, roll, or fly on extraterrestrial environments far better than humans can now or could previously" (Dator 2012, p. 62).

5.5.1.2 Terraforming/Ecological Engineering

Likewise, if terraforming interventions are pursued, these could be kept to a minimum. Terraforming would become a more realistic proposition when the objective is not exclusively linked to the accommodation of humans that mandate stringent habitability requirements. If the goals would become more relaxed, terraforming could serve other purposes such as restoring indigenous life on Mars or other life forms that are better suited to thrive in the planetary conditions of Mars.

Terraforming technologies should be implemented not with the intent to make the planet habitable for human beings only, but to maximize the natural potential of the existing conditions of the planet. In other words, not to re-create Earth landscapes on Mars, but to restore the indigenous landscape—to preserve Mars' uniqueness. Such cooperative efforts would entail re-establishing the atmosphere on Mars based on its natural air composition propensities and reviving whatever life forms that may have existed in its geological history.

5.5.1.3 Virtual Environments

Telerobotics, teleoperations, and remote sensing can serve not only for the purposes of data acquisition but also adaptation to the new habitat. VTs already provide a highly-mediated experience of the landscape through the 'sense organs' of the robotic rover. Here VT can be seen more as an embodiment relation than a hermeneutical relation in that it extends the body's existing perception capabilities.

Virtual environments like CAVE can be used to satisfy both our empathetic as well as our abstractive drives. If they are used to re-create familiar Earth landscapes for nostalgic purposes they can help alleviate loneliness and sense of isolation of expedition members, but on the negative side, if used only for this purpose VTs can also perpetuate feelings of alienation and displacement by inhibiting the full experience of living on Mars. If they are used additionally to re-create Martian landscapes, not only will they impart a more intimate knowledge of their surroundings than just through quantitative data, but will also enhance the ability to experience the environment more fully, which will facilitate stronger empathetic connections to their surroundings through amplified sensory experience of the Martian landscape.

Virtual technologies could conceivably allow Mars inhabitants to feel the texture of its regolith, smell the particles of ferrous dust, visually take in geological features remote to their site, and hear the sound of wind storms in the thin atmosphere. VTs could provide many of the same benefits of cyborg technologies, but the difference is that in VEs the environment is mediated so that it can be experienced by the unmodified body.

5.5.1.4 Robots, Androids, and Artilects

Robots have typically been assigned to perform the dull, dangerous, delicate, or dirty chores that humans either cannot or do not want to do themselves. This not only limits the potential of robotic technology, but relegates them to an undervalued and subservient status. Instead of subjugating our mechanical creations, robots, androids—and eventually artilects—should be invited to share a multitude of experiences, privileges, and duties alongside humans. They could be enlisted as explorers, scientists, technicians, medical aid providers, caretakers, archivists, historians, and companions. The relationship between robots and humans should be developed to be cooperative and complementary, rather than one of 'master' and 'slave.' Embracing their potential begins with first acknowledging our reliance on them and the value they have in our lives, now and in the future. They have already made significant contributions in the exploration of Space, particularly on Mars, and in addition to their scientific capacities they could be essential in providing emotional support and companionship for humans living far away from Earth.

5.5.1.5 Habitat Architecture Technologies

The habitat enclosure should be designed not with the intent of shielding occupants from the environment (phenomenological barrier) but to create interfaces with the landscape (phenomenological engagement) while still offering protection and security (see Sect. 3.6).

5.5.1.6 Biomimicry

The inclusion of plants and animals in the habitat can serve as inspiration for biomimetic design applications. Over eons of evolution, Nature has solved many of the same problems that we are trying to solve in Space (Benyus 2002). For example, the challenge of how to maximize the amount of heat in a concentrated area has a precedent in the silversword—a succulent found at higher altitudes in the

Haleakala crater on Maui, where there is an abundance of sunlight, the air is thin, and the temperature relatively cold. The plant evolved in such a way that took advantage and optimized the unique conditions of its inhospitable environment. The reflective surface of the plant focuses sunlight at its core, generating a micro-climate warm enough to allow the plant to flower (and thus reproduce), as well as to protect it from the searing UV rays.[12] Such a system could be applicable for sites in Space. Examining how organisms in Nature developed strategies to acclimate and thrive in less than hospitable conditions could provide the basis for design strategies beyond Earth as well.

5.5.2 Cooperative Strategies for Social Adaptation

Just as human migration to Space will require new technologies to help us survive and adapt to our new extraterrestrial environment, it will also require the development of new social structures appropriate to supporting a seminal human community in Space. Decision-making protocols for earlier missions were built on the military structure in which the pilots who made up the crew were trained and acculturated; there was a clear and rigid hierarchy of leadership and chain of command. However, as crew composition began to include scientists, engineers, and non-military personnel, the authoritative military style was no longer an appropriate model. As with technologies, new models of social organization must be explored to best insure adaptation in Space. Cooperative approaches are the most sound on many levels, from how we implement our technologies, to how we interact with new environments, as well as how we establish social structures for burgeoning Space communities.

Moreover, Space travel is too expensive, the required technologies too complex, the scope of required knowledge too broad, and the risks too great, to be undertaken as an independent or competitive endeavor, as it was in its inception. Collaboration between agencies such as NASA, JAXA, ESA, and the Russian, Chinese, and Indian Space agencies are already headed in that direction as a matter of practical necessity. This precedent may very well establish the preliminary conditions for cooperation to flourish in Space settlements. Given the small size but increasing diversity of the crew, the indispensable contributions that each of them will be expected to make, and their distance from civilization on Earth, cooperative, democratic-based communities would best insure their chances for success.

What I propose is a 'symbiocracy,' a cooperation-based social model derived from the biological principles of symbiosis and the political principles of democracy. A symbiocracy would be a form of governance which, unlike the U.S. Constitution, would not guarantee the right to individual happiness, but

[12] Lecture at the University of Hawaii School of Architecture by William Kramer on biological precedents and the applicability for architectural design (2007).

would acknowledge the necessity of all living things to collectively realize their greatest potential. Value is placed on the *differences* between entities and not on their individual strength. "The basic rule of the biological and social universe is increase of diversification, heterogeneity, and symbiotization. What survives is not the strongest, but the most symbiotic" (Maruyama 1978a, p. 459). The greater the diversity, the better chance not only to survive but to thrive, since diversification of life has proven to better withstand ecological crises and catastrophes. Tenets of a symbiocracy would include the preservation of diversity, maximization of benefits for the whole as well as for the individual, and mutual dependence over domination. Because symbiocracy does not rely on or encourage structures of competition and adversarialism, it could conceivably reduce causes of war, violence, and oppression. The negotiative process of symbiotization leads to reconciliation, as opposed to annihilation which results from competitive processes.

Unlike current models of democracy which are based on the assumption of equality, symbiocracy is founded on the premise of vitality. Vitality in this context should not be confused with sexual potency or vigor, but understood simply as a condition of being vital or essential. It distinguishes itself from equality in that it is a qualitative condition, one that implies value through indispensability; whereas equality—though often conceived in valuative terms—is fundamentally a quantitative condition that implies nothing more significant than neutrality between entities. Interdependence is also a requirement of vitality, one which is not inherent to the notion of equality. It entails cooperation between entities for mutual benefit.

Democracy is limited only to entities that are deemed to be equal whereas symbiocracy extends to all entities deemed to be vital. All entities that are not considered equal are then arbitrarily excluded from the realm of political rights. Because symbiocracy is not tied to the assumption of equality, it can be applied to all living things—animals, plants, even microbial organisms—as well as the non-living environment. Rights are therefore extended to anything considered vital. While one might argue that microbes are not 'equal' to us in terms of their individual complexity, they are certainly superior to humans in terms of their collective vitality and symbiotic capacity. It may upon first glance seem an absurd proposition to include entities like bacteria in a socio-political system, but within the context of symbiotic relationships which are essential to maintaining the balance of life, it not only makes sense but becomes imperative.

Compare this proposition with current social structures on Earth which are based on Darwinian principles: Survival of the fittest sets the stage for adversarialism where individuals compete against each other for economic, social, or political advantage—an inherently zero-sum game. Darwinism has justified structures of conquest. A select group of winners attain elite status while the vast majority constitutes losers. Whereas kinships foster greater social cohesion, rivalries, on the other hand, create social divisiveness. Though competitive endeavors spur us on to great discoveries, accomplishments, and wondrous physical feats, competitive behaviors have also nurtured a darker side of humanity, one that leads to conflict, domination, violence, and unethical behavior. Competition creates change-amplifying feedback cycles (Maruyama 1963, 1978b), leading to excess consumption and hoarding of resources which in turn deprive others of what they need. Unfettered capitalism fosters societies of insatiable wants and an ethos of envy and dissatisfaction, which then often results in conflict.

Since competition is a source of conflict, are there other ways to create challenges to exploit our potential that are less destructive? Shouldn't we explore other means by which to encourage growth that is not at the expense of others? Isn't our drive to compete—the underlying motive of which is to demonstrate superiority—also incongruous with the notion of equality that is cherished in western culture? And finally, is competition a necessary part of human nature or fundamental to our evolutionary development, or can we move beyond that? Hibbing and Theiss-Morse's research indicates that "human society has evolved selecting for individuals who are 'wary' cooperators. Cooperation is not 'socially learned'—it is hardwired" (Sullivan et al. 2008, p. 351). This conclusion suggests that symbiocracy is not only a viable alternative but is in line with our natural inclinations.

Symbiotic relationships are based on *mutualism*—in which there are gains for both parties in an exchange—but need not rely on a disposition towards *altruism*—the selfless desire to help another without the expectation of gain. Because it can still flourish in conditions of selfishness, symbiocracy is therefore compatible with principles of natural selection. However, social interaction based on mutualism is more beneficial than interaction based on Darwinian principles in that mutualism promotes more complex relationships and networks between entities, and maximizes collective gains. Symbiocratic structures integrates aspects of both democratic and socialist systems by incorporating both the rights of the individual and the responsibilities to the community, by taking into account the welfare of both the individual and the collective. It differs from democratic structures that protect only the rights of each individual without necessarily protecting those of the collective.

To maintain its viability, the concept of democracy must also assume a condition of rationalized altruism, behavior that is inconsistently exhibited at best; "'rational choice' models often provide incomplete or insufficient understanding of human behavior" (Sullivan et al. 2008, p. 350). Therefore, democratic structures require external incentivization in order to be effective. Symbiocratic structures however, need not rely on the principle of human rationality as a basis for right action. Whereas altruistic behavior cannot be justified under natural selection, it presents no inconsistency within the symbiotic paradigm.

"Cooperative societies are self-sustaining, and have adaptive advantages over societies where people act as individuals in a purely selfish manner" (Sullivan et al. 2008, p. 350). Cooperative social arrangements are therefore the best chance for insuring the survival of seminal Space communities. In establishing the structure of the community and the roles of its members, each individual should be assigned some function that is indispensable to the survival of the community as a whole (preserving diversity and vitality); and as the settlement expands, each community should also retain an indispensable function relative to the entire colony as a whole. In addition, there should be redundancy of skills, so that if one party is unable to fulfill its responsibility, another can step in. Similar to the community model at Los Horcones (Robinson 1998), roles should rotate regularly so that all members can develop redundant expertise as well as a diversification of skills in order to be best equipped to handle catastrophe and crisis—as well as to keep stimulated and challenged.

Cooperative structures will have far-reaching effects on how we evolve as a species. Margulis attests that "the process of macroevolution occurs...when formerly independent organisms begin to cooperate in living communes where they can pool their various talents for mutual advantage" (Gardner 2003, p. 196). Symbiotic mergers with foreign environments in Space, as well as fusions with our technologies, naturally raise the question of how we will define what it is to be human in possible futures. To describe ourselves as *Homo sapiens* in the genetic sense, or even as Earthlings in the geographical sense, will no longer be accurate. The technological approach and the specific means by which we adapt to life in Space have implications for our being that should be rigorously examined.

References

Axelrod, R. (1987). *The evolution of cooperation*. Orlando, FL: Harcourt-Brace Jovanovich Publishers.

Benyus, J. M. (2002). *Biomimicry: Innovation inspired by nature*. New York, NY: Harper Collins.

Capra, F. (1996). *The web of life: A new scientific understanding of living systems*. New York, NY: Anchor Books.

Chaisson, E. J. (2001). *Cosmic evolution: The rise of complexity in nature*. Cambridge, MA: Harvard University Press.

Darwin, C. (2004). *The descent of man*. London, England: Penguin Classic Books.

Dator, J. (2012). *Social foundations of human space exploration*. New York, NY: Springer.

Dawkins, R. (2006). *The selfish gene: 30th anniversary edition*. Oxford, England: Oxford University Press.

Devall, B., & Sessions, G. (1985). *Deep ecology: Living as if nature matters*. Salt Lake City, UT: Gibbs Smith.

Gallagher, W. (2001). *The power of place*. Cambridge, MA: Harvard University Press.

Gardner, J. N. (2003). *Biocosm: The new scientific theory of evolution*. Maui, HI: Inner Ocean Publishing.

Goodall, J. (2005). The will to evolve. In M. Smith (Ed.), *Stelarc: The monograph* (pp. 1–31). Cambridge, MA: MIT Press.

Kelly, K. (1994). *Out of control: The new biology of machines, social systems, and the economic world*. New York, NY: Basic Books.

Kropotkin, P. (2006). *Mutual aid: A factor of evolution*. Mineola, NY: Dover Publications.

Lanier, J. (2010). *You are not a gadget*. New York, NY: Random House.

Lovelock, J. (2000). *Gaia: A new look at life on earth*. Oxford, England: Oxford University Press.

Margulis, L. (1998). *Symbiotic planet: A new look at evolution*. New York, NY: Basic Books.

Margulis, L., & Sagan, D. (1997a). *Slanted truths: Essays on Gaia, symbiosis and evolution*. New York, NY: Springer.

Margulis, L., & Sagan, D. (1997b). *Microcosmos: Four billion years of microbial evolution* (1st ed.). Berkeley, CA: University of California Press.

Maruyama, M. (1963). The second cybernetics: Deviation-amplifying mutual causal processes. *American Scientist, 51*(2), 164–179.

Maruyama, M. (1978a). Symbiotization of cultural heterogeneity: Scientific, epistemological, and aesthetic bases. In M. Maruyama & A. Harkins (Eds.), *Cultures of the future*. Berlin, Germany: De Gruyter Mouton.

Maruyama, M. (1978b). Toward human futuristics. In M. Maruyama & A. M. Harkins (Eds.), *Cultures of the future* (pp. 33–59). The Hague, Netherlands: Mouton.

Robinson, J. (1998). Comunidad Los Horcones: Radical behaviorism in Mexico. *Compendio de Artículos publicados por la Comunidad Los Horcones*, 143–153.

Steiner, D. F. R. (2002). *Human ecology: Following nature's lead*. Washington, DC: Island Press.

Sullivan, B. A., Snyder, M., Sullivan, J., & Chapp, C. (2008). The political psychology of cooperation: Synthesis and prospects. In B. Sullivan, M. Snyder, & J. Sullivan (Eds.), *Cooperation: The political psychology of effective human interaction* (pp. 343–358). Hoboken, NJ: Blackwell.

Taylor, M. (1976). *Anarchy and cooperation*. Hoboken, NJ: John Wiley & Sons.

Taylor, M. (1982). *Community, anarchy and liberty*. Cambridge, MA: Cambridge University Press.

Thieman, W., & Palladino, M. (2007). *Introduction to biotechnology*. Upper Saddle River, NJ: Pearson Education Incorporated.

Ward, P. (2001). *Future evolution* (1st ed.). New York, NY: Henry Holt & Co.

Chapter 6
Futures of Human Evolution

6.1 Premise

The isolated, remote, and radically different geographies of Space will have a profound impact on human evolution, not only in the conventional biological sense, but also through the technological mediations that will be indispensable for dwelling in Space. The process of change will shift from slow gradual increments as it is in biology, to a revolutionary, abrupt, full-scale transformation. We should not expect that we will remain human in the conventional sense—but what does it mean then to be 'posthuman'?

If speciation (based on reproductive compatibility) is the threshold of genetic change, then by what measure can we quantify the changes effected by our technologies? Though there are no distinct taxonometric classifications for change in humans induced by technological modification, it has already become evident that our technologies are changing us both in magnitude and at a pace that threatens the traditional notions of what constitutes human being in the general sense as well as challenges how we define our individual identity.

Humans are constantly re-inventing themselves through their technologies in order to explore new possibilities. Migrating to Space will not turn us into cyborgs; it is in becoming cyborgs that we will be able to migrate to Space in the first place. Whether we define ourselves as posthuman comes down to how we see our relationship to our inventions—either as internalized and integral to the human condition or as something externalized that alters it fundamentally.

6.2 From Adaptation to Evolution

It is important first to distinguish between what is meant by evolution, and how it differs from adaptation. Adaptation is the process of adjustments that are undertaken by an organism (both unconsciously and consciously) to become better suited

E.S. Lockard, *Human Migration to Space*, Springer Theses,
DOI 10.1007/978-3-319-05930-3_6,

to the vacillating conditions of its present habitat, or to differing conditions of a new habitat. Adaptation occurs in a state of imbalance—that is, the relationship between subject and environment is unstable. The objective of adaptation, therefore, is to achieve stability and endure. As such, the nature of those processes is dynamic and ephemeral, with a timeframe that is short-term—ranging from instantaneous to the entire lifespan of an organism.

Evolution, on the other hand, is the outcome of those vacillating processes. In the case of natural evolution, it is entrenched change that ultimately becomes a marking characteristic or defining feature of an organism, not merely a marginal or a genetic manipulation. Evolution is evident in structural or morphogenetic transformations that are robust, enduring, and resilient, so though technology may give rise to profound immediate transformations in the living organism, those alterations must also stand the test of time before we can speak of evolution. The nature of change itself is capricious, but evolution is not; it reflects a state of stability which will only undergo modification after many generations, and only if the environment first changes sufficiently to warrant subsequent adjustments. In contrast to adaptation, evolution is situated in a long-term timeframe, occurring over a span of generations (in biological time), millennia (in geological time), or eons (in cosmological time).

Historically, evolution has been understood in a strictly biological sense, referring to genetic variations that emerge and eventually become inherited traits, depending on whether they confer a fitness advantage. A classical example of biological evolution can be demonstrated in the color of prey animals. Less conspicuously-colored individuals in a prey population are likely to be less vulnerable to predators and therefore possess a survival advantage. As those in the population with this variation outlive their more ostentatious counterparts and go on to reproduce, the trait is subsequently inherited by their offspring, thus making them better adapted to their environment. Eventually this trait becomes genetically prevalent throughout the population, becoming a defining feature of a species.

However, with regards to *Homo sapiens* there are also other modes of evolution at work. Besides biological evolution, humans can evolve culturally and technologically. Cultural evolution is the development of a society's collective behavior, mores and values, political systems, education, etc., that are passed down from generation to generation, traits that are therefore acquired rather than inherited. Religion, for example, is a cultural construct that helps to create kinship between members of a population, affiliations which can be beneficial to survivability. As with biological evolution, a society whose culture does not adapt to the challenges of its circumstances will eventually perish. (The cultural practices of nomadic tribes for example, must necessarily be different from stationery societies in order to survive: in search of new habitats whose resources they consume, they often must leave their elderly or infirm behind; otherwise, they put the survival of the entire clan at risk.)

Another mode of evolution is technological.[1] Throughout human history, the advent of more and more sophisticated tools radically transformed human societies. And as our technological capabilities continue to advance at exponential rates under Moore's Law, their transformative influence will also accelerate.[2] In fact, technological evolution may have already surpassed biological evolution as the dominant mode of change for humankind. According to Kurzweil, our species is on the threshold of a new era, one that will be defined by the merger between technology and human intelligence. Machines will become increasingly more intelligent as technology develops at its current rapid pace, to the point where they will exceed human intelligence and be able to not only self-generate but also produce more advanced 'progeny'. In Kurzweil's vision, human beings and technology will no longer exist as separate entities, but will become fused in a symbiotic relationship with one another, propagating intelligence throughout the Universe.

Modes of evolution other than biological evolution would more appropriately be referred to as 'engineered' evolution, a term that captures the teleological and normative character, and distinguishes it from natural processes. Engineered evolution implies that change is intentionally directed not only towards the perpetuation of the species (a point that has been heavily disputed in the scientific community with regards to natural selection), but also towards preferred attributes selected by humans rather than by the environment.

---

[1] A semantic ambiguity arises when referring to technological adaptation and technological evolution. There are three possible interpretations:

1. Technological adaptation = the regulatory use of technology to adapt to the changing environment which pre-empts a physiological response on the part of the organism (which in turn pre-empts genetic change, and hence, biological evolution).
2. Technological evolution (anthropocentric) = following from technological adaptation, the use of technology has induced a permanent morphological (if not genetic) change in the human species; artifice has become integrated into the body such that the body is no longer complete without it; that is, it has become a permanent, indispensable, and resilient defining feature of the human being.
3. Technological evolution (technocentric) = technology is itself always changing, building upon itself, and becoming more complex; is technology evolving to serve its own end—self-perpetuation—and thus becoming an ends in itself rather than a means? Or is it evolving in service to the ultimate evolution of humankind (human evolution as final cause)? Circumstances can also give rise to a phenomenon similar to punctuated equilibrium: "Arthur C. Clarke referred to spaceflight as a 'technological mutation that should not have occurred until the twenty-first century'...Ordinary technological advances are the products of a gradual process of incremental innovation and improvement. Revolutionary technological change may be brought about by social and political movements operating outside conventional market and scientific processes, and then prove to be unsustainable" (Dator 2012, pp. 27–28).

[2] The invention of the internet demonstrates this acceleration in social transformation: in only the last two decades it has enabled instantaneous, non-spatial, global communication and exchange of information, whereas earlier forms of communication and information exchange developed over a period of centuries. As a result, it has become the single most democratizing force in human history; access to information is no longer the exclusive domain of the powerful elite or hegemony. It has also given rise to the creation of online persona—malleable alter egos—that supplement identity.

6.3 The Dual Nature of Human Evolution

Humans can no longer be seen as passive subjects reacting to the external forces of the physical world; we have become both the architect and navigator of those very forces which we may purposefully and selectively manipulate to suit our desire. Human evolution is no longer a natural or passive process but one that has become a deliberate, pre-meditated act. We don't only evolve *after* migrating to new environments as is the case with all other species; we evolve *in order to* migrate. Our transformations are driven by purpose and are teleological in a way that natural biological evolution is not. We evolve through intelligent design, where we ourselves are the designer. From this point of view human evolution in Space has already begun, even though we have yet to set up the first communities on Mars. It began well before we sent the first astronauts to live on the Space stations; it began with our *decision* to explore Space in the first place. Since then we have developed technologies at an alarming rate that have prepared us to eventually migrate beyond Earth, much like previous technological advances allowed us to explore formerly uncharted areas of our home planet, and this technological evolution has already begun to take effect.

Unlike biological adaptation which is reactive in that it occurs *after* exposure to new environmental conditions (post-cursive evolution), adaptation through the introduction of technological inventions are proactive in that they occur *prior* to exposure to new environments (pre-cursive evolution). Technological modification is the antecedent that not only allows us to inhabit new environments, but *compels* us to explore new conditions to which we are not physiologically suited.

How then is human technological evolution measured or defined—and how is it distinct from our biological evolution? Technology is often defined as the products of human invention, as opposed to what is produced by Nature (Kelly 1994). But this distinction between the human-made vs. Nature-made becomes problematic when we consider that technology is merely an extension of natural processes, practiced not only by humans, but also by many other species that preceded us.[3] Kelly argues that animal technologies are produced by instinct (thus deemed 'natural'), and human technologies are produced by the mind (Kelly 2010), based on the unjustified assumption that what we do is *not* instinctual, and what they do is *not* without mental ingenuity. This concept of technology once again cleaves humans from Nature in an arbitrary fashion.

Biotechnologies straddle the categorical fence and blur the distinctions between the two types of evolution. Biological evolution—which has been traditionally viewed as synonymous with natural evolution—is defined as genetic change over

[3] "Margulis and Sagan comment on how modern techno-scientific humanity's first steps in genetic engineering are a return to and appropriation of capacities bacteria have possessed for several billion years: 'Our ability to make new kinds of life can be seen as the newest way in which organic memory—life's recall and activation of the past in the present—becomes more acute'" (B. Clarke 2008, p. 174).

time, and as such does not include changes initiated by technological manipulation. But changes introduced by genetic engineering[4]—which could conceivably be defined as being both biological and technological—would be highly accelerated, producing "major changes within a few hundred years" (Hart 1986, p. 283). But more impressive than the compressed timeframe for extraordinary change is "the probable ability of humans to significantly influence their environments and their genomes so that morphospeciation will no longer be a spontaneous 'natural' event" and that evolutionary potential "would be nearly limitless" (Valentine 1986, p. 273).

Human evolution cannot be comprehensively described by the conventional principles of science; the processes of evolution are evolving themselves. We are not only the subject of evolution but have become the architect of our own transformations. Furthermore, given that humans can evolve through biological, cultural, and technological means, evolution can both precede migration to new ecological niches as well as subsequently respond to them.

6.4 Evolutionary Trajectories for Human Migration to Space

According to the traditional models of evolutionary biology, there are specific conditions that must be present in order to give rise to evolutionary transformation. Those conditions are: (1) substantial change in environmental conditions, (2) geographical isolation, (3) favorable mutations (i.e., those that confer a fitness advantage), and (4) small population clusters in which changes can take hold (Valentine 1986, p. 268). Migration to Space would certainly meet all the criteria for evolutionary transformation: The first pioneering crews to settle in Space will not be large—likely less than 12 people for seminal colonies; their distance from Earth imposes a radical geographical isolation, especially since the possibility of return is unlikely; and exposure to cosmic radiation—one of the primary causes for the generation of genetic mutations here on Earth—will be increased, thus mutations can be expected to be greater and more frequent (Crosby 1986, p. 279), although mutations too far from our somatic capacities pose danger to survival.

Proponents of symbiotic theory such as Margulis, as well as proponents of punctuated equilibrium such as Stephen Jay Gould, argue that the Darwinian account of gradual cumulative change cannot provide the sufficient catalyst required for the process of speciation. Finney writes:

> Gould and other paleontologists maintain that major evolutionary divergence proceeds through bursts of speciation, through the comparatively rapid splitting off of separate lineages from the ancestral stock, and not by the gradual transformation of that stock.

---

[4] This would only apply to germline interventions which are heritable and therefore enduring, and not to somatic interventions which perish along with the subject.

> According to this increasingly accepted view, rapid speciation occurs primarily in very small populations that have become geographically isolated from the ancestral stock. Where genetic change is resisted by large populations well adapted to their environment, favorable genetic mutations can easily gain a foothold in marginal geographic areas where pressures for natural selection can be intense and then spread quickly through the small populations that have become isolated there (Finney 1986, p. 23).

It was precisely these conditions of isolation that gave rise to a number of precursor species of hominids on Earth. But with humans now numbering over seven billion and occupying almost every region on the planet, these conditions have all but disappeared, and *Homo sapiens* as a result have no longer evolved much as a species biologically. Finney states that "[the] once vigorous speciation response that brought forth so many species of Australopithecus and then Homo has now been greatly dampened. Now because we, the surviving hominid species, form one intercommunicating, interbreeding world population, rapid speciation would no longer seem possible...[however,] if our descendants spread far and wide through Space, the forces of evolution now braked on Earth will be released once more" (Finney 1986, p. 23). Valentine makes a similar claim:

> It is sometimes suggested that evolution has run its course in our species. Assuming that galactic colonization occurs, experience with 600 million years of animal life on Earth suggest differently. Experience does suggest that many colonizing populations will probably not change too much biologically, although to human perceptions a minor evolutionary fluctuation might be all too obvious. But a certain percentage of the colonizing populations should undergo significant changes that result in the sorts of morphological differences that characterize different species of hominids. Some human isolates may well undergo such extensive repatterning of their morphology and physiology as to establish entrée into a new adaptive zone (Valentine 1986, pp. 273–274).

Both Finney's and Valentine's claims underscore the twofold nature of evolutionary change: that *precursive* change must occur prior to migration (through the development of technologies that enable us to venture to new habitats); and that *postcursive* change (through biological adaptations) will result once we are exposed to the circumstances of the new environment.

The potential for evolutionary transformation in Space exceeds what we could ever possibly imagine, but whatever direction those changes may take, it is likely that the colonizing populations will no longer bear any resemblance to their Earthly counterparts. Given their conditions of isolation and their small numbers, human diasporas in Space "are practically tailor-made experiments in inducing evolutionary novelty...producing distinctive new morphospecies, the origin of which should involve time scales of only thousands of years..." (Valentine 1986, pp. 272–273). The vastness of Space, according to Valentine, is comparable to the Cambrian period of Earth in which many "new species did more than enter a niche; they entered whole new adaptive zones that happened to be largely empty and underwent extensive morphological repatterning to create novel body plans. Some of those were so distinctive that we recognize them as phyla, classes, and orders, but they originated so rapidly that we cannot resolve their origin in the geologic record" (Valentine 1986, p. 273). He concludes that "within 2 million years the descendants of *Homo sapiens*, scattered across the Galaxy, will exhibit a diversity of form and adaptation that would astound us residents of today's Earth and that even then our evolutionary potential will hardly be scratched" (Valentine 1986, p. 274).

The prospects for human evolution in Space are fascinating from the anthropologist's perspective. But interstellar human migration is considered by some to be 'unnatural,' based on one of two rationales: either it is unnatural because our evolutionary history is rooted on the planet Earth, and therefore venturing into Space is a deviation from our natural evolutionary path; or it is unnatural because human survival beyond the protective confines of Earth necessitates the intensive use of advanced technologies. Chris Hables Gray compares the environment of Space to that of fabricated virtual environments: "Space is an unnatural environment for humans. Unlike cyberspace, which was constructed for and by people so that they could instantiate various fantasies. . .Space has always been there and it is not hospitable" (Gray 2002, p. 138).

While the opponents of Space exploration argue that it is an aberration for humans to live in Space because we evolved in terrestrial conditions—and therefore conclude that we were not meant to live in extraterrestrial environments—the evolutionary history of all biological life would demonstrate otherwise: that it is, in fact, essential for the perpetuation of a species to migrate and expand to new habitats for which they were not originally suited. A diverse group of scholars, such as Ben Finney, Alfred Crosby, Richard Lee, Douglas Schwartz, and Nancy Makepeace Tanner, draw analogies between migrations in human evolutionary history in order to make this point. Finney cites the Polynesian migration across the Pacific Ocean to the islands of Hawaii; Lee examines the Viking expansion throughout Scandinavia; Crosby describes the settlement of the Americas by the Europeans; and Tanner compares the journey from the jungle to the savanna and the resulting transition from apes to humans. "To survive, animals must explore their environments to find sources of food and living space, and a successful species is one that expands its habitat through migration" (Finney 1986, p. 21). Finney brings to our attention that the development of primitive technologies which allowed us to explore greater distances conferred to *Homo sapiens* a fitness advantage over other hominid species. This is the reason why *Homo sapiens* is the only hominid species still left in existence. He also states that "the use of technology to expand beyond Earth would be entirely consonant with the whole trend of human evolution. From the time those most adventuresome of apes left the tropical forest to seek a living in the grasslands of the African savanna, our ancestors have been inventing technology to adapt to new environments and to expand over the globe" (Finney and Jones 1986a, p. 335).

Which brings us to the second objection based on technological dependence: that our prospective migration to the distant and inhospitable environments of Space (these are relative terms, of course) requires intensive technological intervention. But this is also nothing new in human history on Earth. One of the first examples is the development of sacks made from animal hides which allowed early *Homo sapiens* to venture further away from their original habitats with the goods needed for prolonged survival (Tanner 1986, p. 223). If we accept the premise based on our evolutionary history that the development and use of technologies are in fact, what makes us human—as being not just incidental but *essential* to the human condition (Clark; Hayles; Kelly)—then the changes that are induced by those technologies should be likewise framed as a natural extension of our nature. The claim that human migration to Space is somehow unnatural because it requires extensive technologies to enable us to live beyond Earth makes a wholly arbitrary distinction based on degree of intervention. When the early hominids invented cargo pouches that allowed them to travel deeper into the savanna, would that then also be considered unnatural? When the first humans wore animal skins that gave them warmth and allowed them to survive in latitudes further from the equator, would that not also be unnatural? The threshold between Space and Earth may seem to modern-day humans as extreme, and the challenges to survival a formidable deterrent, but it is no more insurmountable than the boundary between the tropics and the savanna was for the early hominids, or the distance between continents separated by oceans was for the early explorers. In each case, human inventive ingenuity made all significant migrations viable.

This premise of technology as intrinsic to human nature also challenges the distinction between natural and artificial evolution. Natural evolution, at least from a Darwinian perspective, is understood as the change that is brought about by an organism's capacity to adapt to environmental conditions; only changes in genetic composition are recognized. But if humans have biologically evolved to become inventors and innovators, then it is arguably a genetic imperative (and thus natural by biological standards) to incorporate technology into the evolutionary process.[5] Instead of the natural/artificial dichotomy which underscores our separation from Nature—a schism that should best be abolished—the notion of what is 'natural' must also evolve accordingly. Both technological and biological evolution are indeed natural.

The prospects for human transformation will not only be in response to the extreme environmental differences that we will confront in Space, but also determined by the emerging technologies that will allow us to venture to and inhabit extraterrestrial sites in the first place. In addition to technological and environmental conditions, two other factors influence evolution: the subjects themselves and the circumstances surrounding their migration—as well as, of course, time.

[5] The use of technology is not uniquely or exclusively human; for example, beavers build dams and birds construct nests. Nature is replete with examples of exquisite and elegant technological solutions that humans cannot even come close to replicating.

6.5 Timeline Projections

6.5.1 Immediate Future (Timeline: 10–20 Years)

6.5.1.1 Circumstances

Within this generation we will see the first humans travel to the Moon and Mars. Private aerospace companies, such as Inspiration Mars, are already poised to send the first couple to orbit (but not land on) Mars in 2018, and there seems to be no shortage of volunteers willing to make that voyage, even given the dangers and high risk of death. NASA has longer range plans to develop the Orion Mars Mission, which will deliver a human crew to Mars in the 2030s with the purpose of establishing the first human outpost there. The astronauts are expected to return to Earth after one year, in another spacecraft sent separately called the Earth Return Vehicle (ERV). Any human population on Mars or the Moon during this period will be of a transitory nature, and their status will be only that of visitor, in which case the need for long-term adaptation is not of great concern.

6.5.1.2 Adaptive Technologies

The types of technologies needed for the goals of this timeframe are already viable, or will be viable within the decade. It is unlikely in the compressed timeframe ahead that any new technologies that will be immediately deployable for Space will emerge; only extant technologies, or those currently in development, will be refined and perfected. Since these technologies have achieved viability, they are relatively easy to forecast:

- Habitat architecture will continue to be designed primarily for the increasing demands of survival under more challenging conditions. The focus will still be on shielding and protection from the dangers of Space, but will incorporate more features for habitability in the way of creature comforts. Efficiency in structure to reduce payload demands will still drive the design; habitat modules will remain spherical or cylindrical, informed by rational geometries that are symmetrical, non-idiosyncratic, and non-organic.
- Cyborg technologies will take the form of external prostheses, such as pressurized suits, gloves, helmets, tools, computers, etc. There are no plans yet to develop advanced cyborg technologies, such as somatic gene therapy, that will help the body acclimate to conditions of Space, and not enough is yet known about the effects of pharmaceuticals on the body in partial or zero gravity conditions. The use of drugs or neurochemicals will require much more extensive research before they can be viable for the purposes of somatic adaptation in Space.

- Because the habitat will be used to shield its occupants from the environment, and because of the extreme distance from Earth, virtual technologies such as telecommunications, teleoperation, and remote sensing will play an even larger role in the immediate future. Existing VE simulations could be easily developed to replicate the Mars landscape. While these currently are limited predominantly to the senses of sight and sound, a new generation of VTs could conceivably be invented that would engage the other senses, which would help to make the experience of living in the Martian environment feel more authentic and less encapsulated.
- Robots, which have already paved the way for humans to settle on Mars, will be a vital component of the outpost but more as an object of service. They will have only limited autonomy and function more like extensions of the operator. However, robots who can perform social functions are already being employed on Earth, and so sociable robots could easily be integrated in a Martian outpost. Robots as companions may seem too frivolous in this initial phase, and might also be met with skepticism by the crew, but as they gain wider acceptance here on Earth, such robots will eventually be embraced for long-duration interplanetary missions.
- Terraforming will not be explored at this time because the technology to make the Martian surface fully habitable for humans will not exist yet; even partial terraforming to make it habitable for plant life which is easier to achieve would still be too expensive and would require too much time to make it a practical option. Issues of feasibility aside, the controversy of using invasive technologies, should any life be discovered on Mars, would discourage any such attempts.

Since long-term adaptation for the first human delegation to Mars will not be a priority in this phase, technologies will continue to be driven by the avoidance approach. This phase, in which travel to Space is still a novelty, will be characterized as a period of excitement and adventure. Space will still be viewed as the unknown and mysterious realm, eliciting feelings of trepidation, anxiety, and alienation. This attitude will prevail until either traveling to Space becomes a more mundane event or the requirements for long-term adaptation are acknowledged and addressed.

6.5.2 *Near Future (Timeline: 20–50 Years)*

6.5.2.1 Circumstances

After the success of the first temporary outposts (assuming there are no tragic setbacks), settlements will be established for permanent habitation on Mars. As they increase in number, these isolated clusters will form small cohesive communities, which may self-organize according to optimal operational size, rotation of occupants, their respective locations on Mars, or their functional role in the larger aggregate. This next generation of Space travelers will not be characterized as visitors or tourists like their predecessors, but as immigrants who will live out the rest of their lives on their new adopted planet. With their change in status, their views of Mars and their relationship to their new surroundings will change as well. Mars will become a second home for this human diaspora. Issues of citizenship and nationality will arise, and a unique Martian culture will begin to develop, including economic, social, and governance structures.

Martian geographies of place, such as 'region,' 'quadrant,' 'parish,' 'vicinity,' etc., will also emerge, and the planet will possess a human history beyond an ephemeral presence. Many of the original challenges of migration will have been solved, though some failures and setbacks will have been inevitably experienced. But as establishing settlements on Mars will become more commonplace and even routine, the initial novelty and thrill of the previous phase will eventually disappear. As extraterrestrial venues become more familiar to us, there will be less anxiety associated with living in Space. However, Earth will still be humanity's frame of reference, and so to some extent anywhere beyond Earth will still be perceived as foreign.

How humans adapt to life on Mars will depend in large part on the reasons they travelled there. If those selected to make the journey are part of an elite class (economically, intellectually, professionally, etc.) then migrating to Mars will be considered a privilege, and this will generate a wider appeal to go there. If Mars becomes a refuge from an Earth that has fallen to environmental disaster, there will also be a strong desire to be chosen to emigrate, but not because of the allure of the Red Planet itself. If the prospect of living on Mars loses its novelty and interest, or if earlier outposts suffer horrific outcomes, it is also possible that the planet may become an exile colony like Australia once was, in which case Mars would no longer be a desirable destination at all. Each scenario frames expectations and perceptions of Mars very differently, and therefore influences implementation of technology.

6.5.2.2 Adaptive Technologies

Technologies that will be viable for this phase of exploration are only feasible but not yet viable at present. They may be in the earliest stages of their development, but will make significant advances in the next few decades. They are still relatively easy to forecast as they already have a foundation upon which to build.

- VE technologies will continue to become more sophisticated, with a higher degree of immersion through engaging all of the senses (not just sight and sound), and incorporating movement (spatial aspects) and resistance (material/ physical aspects). The VEs of the future will feel much more real (perhaps using holography) than they are today.

- Genetic engineering (somatic) would modify the individual's body to be better suited to the environmental conditions of Mars which are unfavorable to Earthlings. This would eliminate much of the burden previously assumed by the habitat to protect its occupants. Some somatic therapies would be targeted specifically to do the following: counter the ill effects of reduced gravity on the body such as muscle atrophy or decalcification and bone loss; develop heightened tolerance for increased exposure to cosmic radiation; acclimate the body to adapt to colder temperatures; alter circadian cycles for entrainment to Martian or lunar diurnal cycles; etc. This may also involve the use of neurochemicals and other pharmaceuticals.
- With the challenges of survival behind them, more 'discretionary' habitat technologies will be incorporated to enhance experience, which will include sports and other playful activities; the driving force behind the design of the habitat will be less about sequestration from the landscape and more about association with it. This shift will significantly improve the phenomenal relationship to the environment.
- In addition to their utilitarian robotic functions, a new generation of androids will be programmed specifically for social interaction; they will become integral members of the community but not yet enjoy the same status as their human companions.
- Two conditions will eventually arise that will increase the probability for terraforming: first, planetary engineering technologies will become more advanced, making partial, if not full, terraforming viable; and second, reconnaissance of Mars will have definitively established whether life exists, or once existed, on the planet. If life had existed on Mars but became extinct, there will be a push to reincarnate those organisms, especially if they are found to be unique to the planet and are unlike life on Earth (for example, possessing right-hand chirality). Terraforming would not be exclusively for the purposes of human habitation, but also to resurrect indigenous Martian life.

With many of the risks of initial habitation minimized, and with more leisure time available to the settlers, sports activities will occupy a more prominent position in life away from Earth. Existing sports may be modified for the different environments of the Moon or Mars, or altogether new games may be invented. The technologies for Martian or lunar sports will not be conceived and developed on Earth, but will arise directly from the experience of local conditions. (It would be impossible to imagine an activity like surfing, or to come up with a design for a surfboard, if one has never experienced the force of waves in the ocean.) It will not be simply a matter of being lighter and able to jump higher in partial gravity; athletic activities cannot be expected to be like on Earth, as the conditions of inertia are different under reduced gravity and thus the body will have less counter-resistance to movement. Earth humans who have only known 1G will not be able to anticipate the experience of those activities.

Unlike the first phase, addressing long-term adaptation criteria will be critical in the following generation of Space settlers. Practical technologies will not be sufficient and should be supplemented with technologies that also amplify experience of the Martian landscape. Only through enabling engagement with the physical surroundings will the inhabitants develop a sense of belonging and a feeling of rootedness. However, the approach that will be adopted to implement those technologies is not foreseeable. We will come to a fork in the road where we must decide whether we subscribe to either a subordination, domination, or integration approach. This will not be determined solely by the innate properties of the technologies themselves, but also by the intent with which we implement them.

6.5.3 Distant Future (Timeline: 50+ Years)

6.5.3.1 Circumstances

Some inhabitants of Mars within this timeframe will either have been very young when they arrived there (with little or no memory of life on Earth) or were born on Mars, making them the first "kama'aina" or indigenous human Martians. To those who are native, Mars will be seen only as home; comparisons with Earth will no longer be made as Earth will no longer be a point of reference—in fact, everything will be compared to Mars as a *new* frame of reference. Perceptions of the planet as hostile, merciless, unforgiving, or treacherous will disappear with this generation. Mars will not seem exotic in the least. Their birthplace will be their benchmark of what will be familiar or foreign to them, as well as be the center point for their circle of empathy. Life on Mars will be as mundane and banal as growing up in a small, remote, sparsely populated town in Alaska is today. Population clusters will be less isolated and more widespread than in the previous phase.

6.5.3.2 Adaptive Technologies

Technologies for the far future are only speculative, and therefore most difficult to forecast. Any attempt to describe them will appear at best disingenuous. Given the accelerated rate of the development of GRIN technologies, we can expect that the kind of inventions that will exist in this time will be inconceivable today, much like an adult in the 1950s would never have been able to imagine a smart phone or iPod. That said, one can at least map out possibilities from what is known today:

- Germline therapies—which stir heated debate at present—will have become more acceptable, as they will be the most practical alternative for the reproduction of offspring that are adequately suited to the conditions of the Space environment and will better insure their survival. Conventional pregnancy (fetal gestation and birth) may prove to be too difficult or dangerous in reduced gravity, in which case children will be conceived outside the womb.

- As humans become more integrated with their technologies, and robots are designed to become more like humans, the ontological gap between them will diminish. Artilects—robots who possess complete autonomy and much more sophisticated social and intellectual skills—will be capable of engaging in all the same types of relationships that humans enjoy with each other. They will be recognized as synthetic life forms, commensurately endowed with the same rights and status as humans.
- Whether proactive terraforming strategies are pursued or not, there will be a radical transformation of the planetary ecology by virtue of the mere presence of humans. The degree to which directed terraforming will be required will depend on the degree of cyborg interventions that may render many environmental modifications unnecessary.
- Like artilects, virtual technologies will become so sophisticated that they will no longer be characterized as virtual. For instance, if in the earlier example of the virtual fireplace one can see the flickering light of the fire, smell the wood burning, hear the crackling of the flames, feel the heat given off by it, and get burned by touching it, then how would the experience between the real and the virtual hearth be any different? The differences will continue to diminish to the point where boundaries between what is virtual and what is real become negligible, and the term 'virtual' will be rendered meaningless.
- Habitat architecture will become biosynthetic, combining nanotechnology in materials and biomimetic design processes to create biogenerative structures—buildings that are like living organisms in that they can be grown (instead of being built), adapt and evolve according to changing environmental conditions and occupant needs (unlike architecture of the past which was static), and heal itself (rather than be repaired). Buildings will not reflect a rigid rational order imposed on them by its human designer. The morphology of structure will be born of natural processes, much in the same way that living organisms develop their forms. Building will become truly organic, responsive to idiosyncratic conditions of its unique surroundings, and customized to its particular occupants (Fig. 6.1). The boundaries between what is living and what is non-living will become blurred even further and may at some point become indistinguishable. Biosynthetic architecture will become the ultimate adaptive mediation between the occupant it interiorizes and the surroundings it exteriorizes.

Humans will not reach this phase unless they have been successful in fully adapting to their non-Earth habitat. This phase culminates in the ultimate integration between human life and the environment around it.

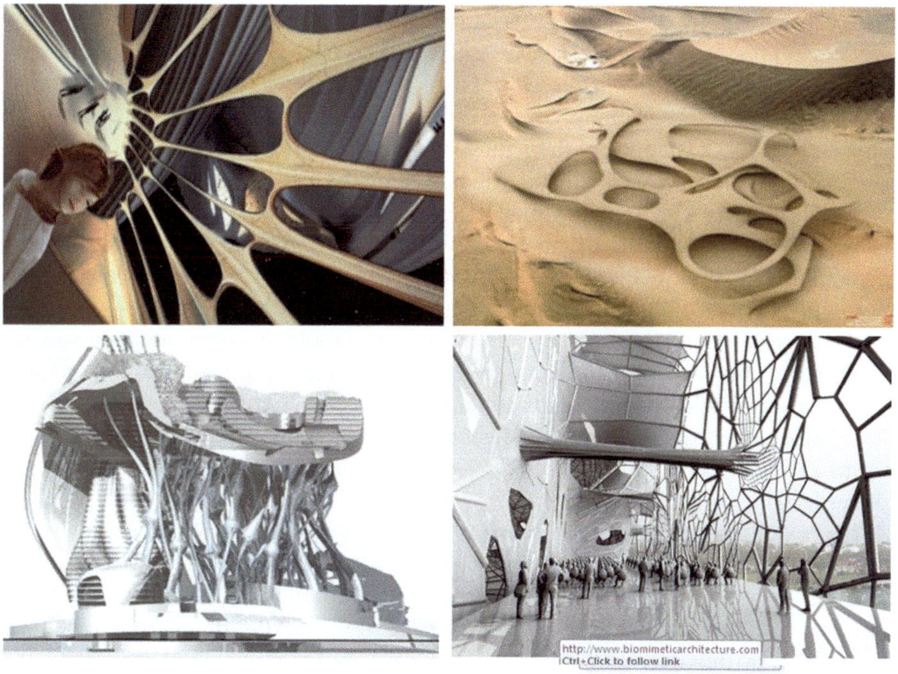

Fig. 6.1 Biomimetic structures (images courtesy of www.biomimeticarchitecture.com)

6.5.4 The Consequences of Space Technologies for Human Being

Virtual reality, terraforming, robotics, and the habitat architecture, all involve technologies that either act as an intermediary between subject and environment or modify the environment directly, minimizing the burden on the human inhabitants to make the necessary physiological adjustments in order to adapt to the radically different conditions of Space. In contrast, the locus of change for cyborg technologies resides entirely within the human subject. However, not all cyborg technologies have the same impact.

In full prosthetic regalia, Space inhabitants would constitute something other than what we would call conventionally human. Typically though, prostheses can be removed and the body restored back to its former condition; thus a return to Earth and re-assimilation with Earth-humans would still be relatively easy. On the other hand, if embedded implants cannot be removed, and its effects are irreversible, then a person may not be able to re-assimilate into human Earth civilization. In either case, humans would still remain human strictly in the biological sense, since pre-emptive interventions would need not entail genetic change, but they would be perceived very differently than humans on Earth.

Unlike prosthetic devices, pharmaceuticals and neurochemicals affect the brain, which can change mood and behavior. Body chemistry is very much tied up with personality, which is one of the aspects we most strongly associate with identity. Again here, although genetic composition is not altered, substantial changes in personality reflect changes in the nature of individual identity.

Genetic engineering will in fact change humans on a biological level. While alterations in the genome may make humans better suited for Space travel, if irreversible, these modified humans may not be able to return and assimilate to the conditions of Earth again. They would no longer be genetically similar, or even biologically compatible with Earth humans. These Space-faring entities—no longer what we would call human in natural terms—would become another species altogether. This abrupt and compressed evolutionary process through genetic manipulation would constitute an artificially-induced punctuated equilibrium, similar to symbiogenesis in that a transfusion of new or foreign genetic material gives rise to a hybrid, but novel form of life—a derivative of the former organism, but no longer bearing any resemblance to it.

The extensive changes that humans will undergo by virtue of technologies necessary for living in Space, as well as how they will alter relations with humans on Earth, push the envelope of identity and call into question what it means to be human. We cannot apprehend what the humans of the next millennium will be like, no more than our ancestors from the previous millennium could possibly have imagined its present-day progeny. Aside from the ontological questions, there are also valuative issues that arise from these changes. Some argue that cyborgian technologies will not improve humanity, only debase it (Fukuyama; Joy; McKibben), while others believe that humans have always been cyborgs—and it is precisely the relationships with externalities that define the human condition (Clark; Hayles; Stone; Haraway).

6.6 Human Being and Becoming

As the relationship between humans and their technologies becomes increasingly enmeshed, boundaries between the organic and the mechanical continue to dissolve, eventually leading to the literal and figurative decomposition of the liberal humanist's conception of the human being as a distinct, separate, and autonomous entity. The relationship between a subject and its environment is likewise not clearly distinguishable as was once thought to be. Human being is not a whole or wholesome entity; it is adulterated, impure, distilled, fused, fragmented, intertwined—comprised of symbiotic couplings that incorporate the aspect of otherness into the self by enveloping external entities.

Whether human being is understood as a relation between the mortal and the divine, or between the mind and the body, the ontological discourse has historically been framed in dueling dualities. As each new dichotomy emerges it renders the previous one irrelevant—while still perpetuating shifting ontologies of humanness as binary instantiations. The concept of the posthuman however, is challenging not only our ontological status, but the tendency to define human identity in exclusive propositions.

In resisting defining humanness in dualities that are pitted against one another (a tendency that is reinforced by Darwinism), there is a new temptation to consolidate disparate components of identity into a holistic, unified entity which is equally reductivistic. The posthuman conceptualization does not give in to this temptation, acknowledging the multiplicitous—and even contradictory—nature of human being. Characterized by plurality rather than singularity, a person is understood not as a rational, indivisible whole, but rather as a complex, interdependent assemblage that can also embrace otherness (DeLanda 2006). In her *Cyborg Manifesto*, Donna Haraway writes that "cyborg imagery can suggest a way out of the maze of dualisms in which we have explained our bodies and our tools to ourselves" (Haraway 1991, p. 181). Instead of "endless splitting and searches for a new essential unity. . .there has also been a growing recognition of another response through coalition—affinity, not identity" (Haraway 1991, p. 155). Multiplicity is, in fact, characteristic in all living organisms: "On the level of biotic systems, Margulis and Sagan have been eloquently arguing for over two decades about the natural metamorphoses of biological evolution: 'the discrete merger of separate systems into hybrid consortiums is the way the world works'" (B. Clarke 2008, pp. 194–195).

6.6.1 *Characteristics of the Posthuman*

There are several common themes that can be distilled from the literature on posthumanism:

6.6.1.1 Embodiment

The first premise is that the posthuman identity requires material instantiation. Andy Clark, who argues that to be human is in fact to be a cyborg, states: "embodiment is *essential* but *negotiable*. Humans are never disembodied intelligences" (Clark 2003, p. 114). Contrary to the liberal humanist model however, embodiment is not limited to our biological packaging: "It is a mistake to posit a biologically fixed 'human nature' with a simple wrap-around of tools and culture; the tools and culture are indeed as much determiners of our nature as products of it. . .[Our brains are naturally predisposed to] the recruitment and exploitation of non-biological props and scaffolds (Clark 2003, p. 86). . .[which] can quite readily project feeling and sensation beyond the biological shell" (Clark 2003, p. 62). Instead of humanness based on essences, "the human person emerges as a shifting matrix of biological and non-biological parts. . .it is our basic *human* nature to annex, exploit, and incorporate non-biological stuff" (Clark 2003, p. 198). Haraway adopts a similar position when she writes in her *Cyborg Manifesto*, "the machine is us, our processes, an aspect of our embodiment" (Haraway 1991, p. 180).

The second premise is that physical embodiment—again in contrast to the notions of the body in classic humanist thought—in no way implies defined boundaries or distinctions from either mind or environment. According to Clark, "...our sense of own bodily limits and bodily presence is not fixed and immovable. Instead, it is an ongoing construct, open to rapid influence by tricks and by new technologies" (Clark 2003, p. 59). "The image of the physical body with which we so readily align our pains and pleasures is highly negotiable. It is a mental construct, open to continual renewal and reconfiguration" (Clark 2003, p. 61). Pepperell also makes the case that "since the human cannot be separated from its environment for any length of time without coming to harm...it seems the human is a 'fuzzy-edged' entity that is profoundly dependent [on] its surroundings, much as the brain is dependent on the body. We have only to consider the perpetual exchange of liquids, chemicals, and energies in the form of urine, feces, menstrual fluid, hair, air, sperm, food, water, skin, sound, light, and heat, to recognize how deeply integrated into the environment we are" (Pepperell 2003, p. 20).

6.6.1.2 Consciousness

First, consciousness is not *central* to identity in the posthuman view: it is not a primordial condition of human being, but rather an emergent feature generated from rudimentary processes that complexified with evolution to eventually give rise to self-awareness and other higher-order mental attributes. Hayles goes even further, relegating consciousness to a subordinate, and arguably inessential role in the reconceptualization of the human being: "The posthuman view considers consciousness, regarded as the seat of human identity...as an epiphenomenon, as an evolutionary upstart trying to claim that it is the whole show when in actuality it is only a minor sideshow" (Hayles 1999, pp. 2–3).

Secondly, consciousness is not *centralized*: like cognition it is dispersed, and thus is not considered to be the locus of human intelligence. Pepperell describes the ambiguity of such delineations: "an integrated continuum exists throughout consciousness, body and environment such that any distinction in that continuum...is necessarily contingent and arbitrary" (Pepperell 2003, p. 22). Catherine Malabou also informs us that even within the organ of the brain there is no seat of consciousness, but rather a network, or 'assemblies' of neural connections: "Cerebral organization presupposes the connection of neurons in networks...[in which] there cannot be, by definition, a privileged vantage point. The network approach is necessarily local, never centralized or centralizing" (Malabou 2008, p. 42).

In dissolving the mind/body dichotomy, subjectivity has become equivocal as a result; our physical boundaries are not defined by our biological substrate, nor are our mental faculties contained within our brain or limited to our bodies. Identity is nebulous in that it expands and contracts, making it indistinguishable from environment. And since the posthuman is no longer demarcated by its biological boundaries, it makes no sense to speak of attributes such as free will, independent agency, autonomy, or centralized control, that have historically comprised the liberal humanist subject.

6.6.1.3 Cognition and Intelligence

Cognitive processes do not reside exclusively within the human subject but are distributed to both animate and non-animate agents that—taken as a whole in a multi-systemic, information-flowing feedback loop—comprise intelligence. It is a collectivizing faculty rather than an individualizing one. Hayles emphasizes this point on several occasions: There is no such thing as independent and autonomous agency, desire, or will distinguished from that of 'others,' because "posthuman's collective heterogeneous quality implies a distributed cognition located in disparate parts that may be in only tenuous communication with one another" (Hayles 1999, pp. 3–4). "We participate in systems whose total cognitive capacity exceeds our individual knowledge…'thinking' is done by both human and non-human actors" (Hayles 1999, pp. 289–290). In conclusion, she claims that "the construction of distributed cognition, a construction that has been ongoing for thousands of years… [changed] the relation of human subjectivity to its environment. No longer is human will seen as the source from which emanates the mastery necessary to dominate and control the environment" (Hayles 1999, p. 290).

Pepperell regards intelligence, historically understood as contained within the individual, "as much a function of an organism's interaction with the environment as its brain processing capacity (Pepperell 2003, pp. 17–18)…in posthuman terms, the process of thought is distributed, not only throughout the brain but through the whole body and even into the environment" (Pepperell 2003, p. 96). The dislocation of intelligence from physical presence therefore denies the absolute body as well as objective sense of place, because identity is not only a construct, but one that is also always under construction. In this sense, identity can never be fully or completely constructed, and thus can never be whole. "The notion of a real, central…self is a profound mistake…that blinds us to our real nature and leads us to radically undervalue and misconceive the roles of context, culture, environment, and technology in the constitution of individual human persons" (Clark 2003, p. 139).

6.6.1.4 Multiplicity/Hybridity

Identity as per western tradition has been synonymous with the individual—autonomous, singular, organically whole and pure. In cyborg imagery the aggregates that make up identity cannot be reduced to a unified entity of homogeneous character; they are by nature fractured, a multiplicitous compilation reconstructed from fragments of Other. Bruce Clarke defines the posthuman as "an amalgam, a collection of heterogeneous components, a material-informatic entity whose boundaries undergo continuous construction and reconstruction" (B. Clarke 2008, p. 3).

Haraway refers to cyborg identity as "a potent subjectivity synthesized from fusions of outside identities" (Haraway 1991, p. 74). "A cyborg body is not innocent; it was not born in a garden; it does not seek unitary identity and so generate antagonistic dualisms without end...[nor is there a] drive in cyborgs to produce total theory, but there is an intimate experience of boundaries, their construction and deconstruction" (Haraway 1991, p. 181). Cyborg imagery boldly asserts an illegitimacy, an impurity that we must revel in, in order to overcome the dualisms that underscore regimes of hegemonic domination: "The cyborg is a creature in a post-gender world; it has no truck with...seductions to organic wholeness through a final appropriation of all the powers of the parts into a higher unity" (Haraway 1991, p. 150).

6.6.2 The Epistemological Perspective

The epistemological perspective re-examines the shifting definition of what it means to be a human being, which denotes a fixed state subject to new interpretations rather than a process of change. It speaks to the present but references history; it is concerned with the changing perceptions of what constitutes the state of humanness. In this case, 'posthuman' does not necessarily refer to the subject itself as changing but rather our understanding of it, alluding to an epistemological reconceptualization.

This perspective in the discourse on the posthuman condition explores the reconfiguration of human identity in contrast to that of the Enlightenment, in which rationality, autonomy, and independence defined the liberal human subject. The possession of consciousness, free will, and the intellect arose as uniquely anthropocentric attributes in the wake of the Scientific Revolution. This worldview, in which humans did not inhabit Nature so much as occupy it, persisted for centuries, but has been recently challenged by advances in the cognitive and neurological sciences, in the field of cybernetics, and in research on AI and AL that depicts the human condition very differently from the Cartesian portrait. How technology potentially alters human nature is one integral aspect of the discourse, but is not exclusive to it. The discourse also addresses the larger, more encompassing metaphysical questions of human nature and subjectivity. Though the term itself insinuates a challenge to the concept of humanness, the posthuman subject can still be posited as ontologically human, only redefined; it is rather our understanding of what constitutes humanness that has been reformulated. "The posthuman does not necessitate the obsolescence of the human; it does not represent an evolution or devolution of the human. Rather it participates in re-distributions of difference and identity" (Halberstam and Livingston 1995, p. 10). The focus is on epistemological questions rather than on ontological issues.

Posthuman theory is a radical departure from the traditional ontological model which arose during the Enlightenment. Newtonian physics—in liberating the intellect from religious indoctrination and displacing faith with scientific knowledge—imposed an isolation, a disconnection, of humans from their environment, which in turn informed not only the way we developed and applied our technologies as mechanisms of domination over Nature, but also the development of our social regimes. "We have extended this fragmented view to our human society, dividing it into different nations, races, religious and political groups. The belief that all these fragments—in ourselves, in our environment, and in our society—are really separate has alienated us from Nature and from our fellow human beings and thus has diminished us. . .identity, individuality, and autonomy do not imply separateness and independence (Margulis and Sagan)" (Capra 1996, p. 296). Whereas in the liberal humanist view being human meant being elevated above Nature, in the posthuman view being elevated above Nature dehumanizes us.

The concept of the posthuman asserts that boundaries of self are not fixed and discrete, but rather are fluid and fuzzy. As with intelligence, which relies on the participation and appropriation of external agents, consciousness associated with individual identity does not reside solely in the brain but is dispersed throughout the body and beyond as well. From the posthuman perspective, identity cannot be definitive in that it constantly expands and contracts, making it indistinguishable from environment. Subjectivity is therefore better understood as embedded in an ongoing process of construction and reconstruction with the environment rather than as an indelibly demarcated state of materiality or consciousness.

"Humanists saw themselves as distinct beings, in an antagonistic relationship with their surroundings. Posthumans, on the other hand, regard their own being as embodied [not only] in an extended technological world" (Pepperell 2003, p. iv), but also in an extended natural world. We exalted ourselves as rational beings and came to see the human race as masters of the Universe—reflecting a mindset of domination that led us to use technology to control and exploit our environments. But bound by this view of ourselves we can only feel alienation from that which we have conquered. As long as our technologies are conceived of and utilized as tools of conquest they will only amplify our sense of psychical and physical displacement.

6.6.3 The Ontological Perspective

The ontological perspective looks at the concept of the posthuman as one of becoming, implying a process towards a state we have not yet reached; in this view "post-" simply stands for "what comes after." Implicit in this definition is a line cast out to multiple futures, and therefore has a speculative nature. The discourse is centered on the development of radicalizing technologies, the range and extent of their applications, and the potential consequences for humankind. There are a number of posthuman alternatives that reflect the influence of technology: the 'superhuman' (in which we are still human, only an improved and upgraded version), the 'inhuman' (in which humans are so transmogrified by our technologies that we are no longer recognizable as human—in other words, monsters), and the 'non-human' (in which we are so transformed that we become something different altogether—e.g., another species, but still descended from humans).[6]

Where the two perspectives in posthuman discourse fundamentally differ in their approach underlies an irony: In the discourse framed within the ontological sphere (the question of what humans are becoming), there is the assumption that humankind remained in a relatively fixed and rigid state until the advent of radicalizing technologies, the changes brought upon by which now threaten to eradicate our essential nature. Conversely, in the discourse framed within the epistemological sphere (the question of what it means to be human being), there is the reverse assumption that humankind has always been in flux, and it is precisely this ongoing metamorphosis that is our true, essential, unwavering nature. The aspect of evolution in the ontological view resides in shifting embodiment; in the epistemological view, it resides in shifting subjectivity. Thus the question of whether we will remain the "same" or become something "other" is irrelevant, a quibble over terminology, as we have always been in the process of becoming rather than in a state of being. In positing a definition of human nature as one of capacities, both actual and those not yet realized, the concept of human being reflects a plasticity that can both absorb as well as reconfigure itself, without the threat of self-annihilation that is a pervasive theme in posthuman pop culture.

Andy Clark argues the same position (though with some semantic differences) when he writes that "My sense of myself. . .is conditioned by my understanding of my own capacities and potentials. . .[technological] extensions should not be thought of as rendering us in any way posthuman; not because they are not deeply transformative but because we humans are naturally designed to be the subjects of just such repeated transformations!" (Clark 2003, p. 142). "What could unite us in the future would be our common participation in the fluid, self-directed process rather than any transitory similarities in form. Seen in this light, strange as future humans may become if germline manipulation achieves its promise, they will still remain human" (Stock 2002, p. 183). It is the processes of resistance, destruction, adaptation, negotiation, generation, and transformation that make us human, regardless of what the resulting outcomes of those processes are. Hence, the technological transmutations we are currently undergoing lead to an extension of humanity rather than to its extinction. In this view, our technologies supplement human identity rather than supplant it.

[6] Another possibility which has not received as much attention is that we may not change at all in any meaningful or substantive way, or that we even become biologically extinct—meeting the same fate as over 99.99% of the species have in the history of life on Earth (Margulis and Sagan 1997b, p. 66).

The term 'cyborg' is a staple of the posthuman lexicon, and though it has a wide range of interpretations, it is usually synonymous with an entity that is part human, part machine. This composite entity straddles the three ontological possibilities of human technological development (superhuman, inhuman, nonhuman), and its interpretation depends on how the relationships with embodied technologies are viewed: Are the machinic components parasitical intrusions that deform the body? Are they simply props that are worn like clothing, posing no threat to the sanctity of the body? Or do they become integral parts of the body, inorganic outgrowths, which become symbiotic components that comprise the self?

6.6.4 The Human as Cyborg: Technologies of Penetration and Extension

The merger of the human body with its technologies—mechanical insinuations of the corporeal domain—has more than anything else in human history challenged our beliefs in what it means to be human. Our organic bodies have been traditionally thought of as the receptacle of our consciousness and the material demarcations of our selfhood. As those boundaries become increasingly infiltrated by our technologies—through prostheses, implants, pharmaceuticals, and genetic alterations—the inviolability of the human subject as we have historically understood it becomes eroded. The image of the dystopian cyborg in contemporary science fiction—usually depicted as a human body infested with technological parasites (Locutus from Star Trek) or a robot deceptively cloaked in human skin in order to pass as one of 'us' (the T-1000 from *The Terminator*, Data from *Star Trek*, Bishop from *Aliens*)—attests to the ambivalence we feel towards our advancing technologies. Our collective uneasiness is due to both the speed at which technology is innovating and the nature of these technologies themselves.

All of our utilitarian inventions are in fact, prosthetic implements that augment both our physical and mental capabilities, and thus expand our sphere of action beyond our biological substrate. There are two classes of cyborg technologies: devices that extend and devices that penetrate. Examples of the former include the shovel which extends our physical capability, the telescope which extends our sensory capability, and the calculator which extends our mental capability. Through these extensions, our tools and machines actually become a part of who we are—even to the point of rendering parts of our physical body irrelevant. Stephen Hawking illustrates this through his relationship with his mechanical voice-box: without his communicative devices that allow him to convey his ideas, to share his knowledge, to transfer information, he would not be the brilliant scientist that he is; he would be compromised, constricted from reaching his fullest potential (Stone 1996). Clearly then, the device that gives him speech is as much a part of him as the non-functioning vocal cords he was born with—and arguably even more so because removal of his mechanical organ would have a far greater impact than the loss of his useless biological counterpart. In being external to the body, technologies of extension are viewed as supplementations to our physical contour, and therefore are not perceived as an infringement on the human body.

Technologies of penetration on the other hand—such as cochlear implants, prosthetic eyes, pharmaceuticals, pacemakers, organ transplants, or genetic manipulations—differ in that they move inward from the surface boundary of our biological substrate. Unlike externalized devices which are visibly distinct, detachable, and non-invasive, technologies of penetration differ in that they are: (1) invasive by their very nature; (2) are not visible (and therefore cannot be distinguished); and (3) fused with the body and thus not readily removable. Implants and other embedded objects are perceived to be desecrations of the body, and therefore are more threatening to our sense of humanness than those that extend from the body.

But on a phenomenological level, there is no distinction between interventions that extend and those that penetrate. They both elicit a condition that crosses, blurs, or erases boundaries; they both alter and augment our capacities. There is very little difference between a pair of athletic shoes, a prosthetic running blade, or performance-enhancing steroids; each extends the capability of the individual who incorporates it. But while society is more accepting of running shoes, it is less comfortable with prosthetic racing blades, and is vehemently opposed to pharmaceutical interventions. The ethical controversies surrounding their respective uses are derived from false or artificial distinctions: the former two are externalized, while the latter is merely internalized. But as we incorporate more technologies of penetration, these distinctions will ultimately become indefensible.

To be sure, we have never been non-technological; human history is defined by our inventions as much as anything else. Humankind has always sought to advance its capabilities through the use of tools and devices. But we are perhaps only now realizing that there is a "deep intimacy of agents and their cognitive scaffoldings" (Clark 2003, p. 141) that goes well beyond utilitarian purposes. Our technological bedfellows have insidiously and insistently pervaded our domain of agency.

In the traditional humanist view, prosthetic devices are thought of as artificial intrusions, as something *other* than belonging to the human body proper that restores or normalizes an imperfect morphology. This reflects a relation of exteriority because the human subject is understood as whole, complete, and independent of its environment. Consistent with this view, Clynes and Kline framed their concept of the cyborg in purely biological terms; they believed the proposed augmentations only increased capacities for adaptability, but in no way changed the nature or the essence of human being. For them, there was no relationship between new capabilities and how they shape human existence, because *genetically* we would still be the same.

In the posthuman view, however, prosthesis is no longer understood in conventional terms only as a replacement for a missing appendage, or as a device attached to the body in order to restore its normal functions, but as something that taps into our potential and enhances our existing capacities. The posthuman relationship is understood as a merger in which identity is shared, expanded, and dispersed, as well as one from which novel properties can emerge. It posits the relationship as an enriching one, rather than one of degradation or compromise. "As a trope that has flourished in a recent and varied literature concerning human-technology interfaces, 'technology as prosthesis' attempts to describe the joining of materials, naturalizations, excorporations, and semiotic transfer that also go far beyond the definition of replacement of a missing part" (Jain 2006, p. 19). Prosthetics are not considered foreign to the human body; they do not constitute something external to, or distinct from it. Rather, it becomes another component by which the self is continually reconstructed. Precisely in extending capabilities, the 'other' has become part of the 'self.' The prosthetic aesthetic reinterprets the relationship between aesthetics, the body, and technology as an a priori prosthetic one (Gray 1995), which is reflected in the following passage:

> The posthuman conception of technology is that of an *extension* to human existence, not of an external agent with a separate history and future. . .humans cannot be understood in isolation from the technological environment that sustains them. What makes us human is our wider technological domain, just as much as our genetic code or natural environment. Throughout history, we have sought to distribute our selves, our consciousness, and our intelligence by a variety of means, including language, art, gesture, and music, by encoding the content of our minds in some material substrate, and to extend our physical abilities with tools. This 'extensionist' view of human nature, in contrast to the humanist view, does not therefore make a distinction between the biological substrate of the human frame (what is most often referred to as the human) and the wider material domain in which we exist (Pepperell 2003, p. 252).

In portraying the relationship between the body and technology as a fusion that constructs a new hybrid entity—the cyborg—the posthuman conception is aligned with evolutionary theories of symbiogenesis that collapse the boundaries between self and other. "Entering into a symbiotic and parasitic relationship with remote others. . .the body 'becomes not only a host for miniaturized technological components but also a body of multiple agencies remotely interacting with it, a body that has a much more fluid sense of self, not so much a split of self from body, but rather a self that is extruded'" (J. Clarke 2005, p. 200). When Clark writes that our brains "make us humans especially open to processes of deep biotechnological symbiosis" (Clark 2003, pp. 62–63), he means that our technological scaffolding is an indispensable part of who we are—that is, an integral part of our nature and not a threat to it. Human identity is redefined in the wake of both externalized mergers with prosthetic devices that extend and decentralize embodiment, as well as internalized mergers with implants, nanotechnologies, and genetic manipulations that penetrate and invisibly reconfigure the biological substrate from within. To be a cyborg is merely to acknowledge that all our relationships and all our interactions—with technology, with the environment, and with other living things—are symbiotic in nature.

The symbiotic perspective also informs the processes of adaptation in an escalating technological world that increasingly envelops and infiltrates us. All cybernetic systems are autopoietic; that is, they possess a reflexive capacity to generate itself. This autopoietic process stems from biologist Francisco Varela's concept of enaction, which examines the ways in which systems can evolve within its environment.[7] "Enaction sees the active engagement of an organism with the environment as the cornerstone of the organism's development" (Hayles 1999, p. 155).

In her essay, 'The Will to Evolve,' Jane Goodall claims that "we have brought ourselves to an evolutionary crisis point by generating a technological environment to which we cannot effectively adapt as a purely biological species...yet at this very crisis point [we] may discover a radically new evolutionary direction. Competition, the central driver of Darwinian evolution, is not involved here, since the direction is triggered through multi-lateral fusions. Technology, symbiotically attached and implanted to the body, creates a new evolutionary synthesis, creates a new hybrid human" (M. Smith 2005, pp. 2–4). What that composite entity connotes is open to wide and varied interpretation, straddling three ontological possibilities of human existence: superhuman, nonhuman, and inhuman.[8]

Although in popular culture the cybernetic organism is stereotypically depicted by such characters as the Terminator or Robocop, the cyborg doesn't only represent the merger of human and machine. In the broadest interpretation, a cyborg is an instantiation of any entity with questionable boundaries and elusive distinctions, comprised of components that inform each other. For Haraway, the cyborg can be the product of the fusion not only between self-controlled, self-governing machines (automatons) and organisms—especially humans (models of autonomy), but also between animals (or other organisms) and humans (Haraway 1991). In all the various interpretations however, the cyborg subject is commonly understood as a 'spliced' entity (as in cyb/org), in which plural components are fused together to produce a new heterogeneous, multiplicitous identity—rather than a 'hyphenated' entity (as in cyb-org) where in its attempts at unification and wholeness, the dialectical schism between polarized components is salient (Hayles 1999, pp. 115–116). Where the splice reinforces the conjunctive character of the composite, the hyphen underscores its disjunctive character. The splice is both a metaphorical as well as de facto representation of the compatibility of organisms produced through symbiogenesis, whereas the hyphen is representative of the adversarialism between organisms in natural selection. Hyphenation preserves

[7] The neurological research Catherine Malabou cites in her book, *What Should We Do with our Brains?*, goes to support Varela's theory of enaction. She describes the brain as plastic rather than flexible. By plastic, she means that it has "the capacity to receive form...[as well as] the capacity to give form." (Malabou 2008, p. 5) Flexibility, she argues, allows only for the ability to accommodate, to submit; but the brain is not merely an acquiescent organ that simply responds to and is imprinted by external phenomena—it is also generative; that is, it "co-occurs with a radical modification of the economic and social environment" (Malabou 2008, p. xii). This new model rejects the rigidity and passivity that has been associated with earlier research on the brain. "Plasticity, far from producing a mirror image of the world, is a form of another possible world." (Malabou 2008, p. 80)

[8] To the degree that the ontological distinctions are a matter of interpretation reflects the co-dependence and co-evolution of both the evolving materialization as well as the ongoing narrative of the posthuman subject (hence also an autopoietic entity); one therefore cannot be understood without examining the other as well.

and perpetuates the tension between dissimilar elements. It assumes an a priori alienation and polarizes through asserting a status of relative 'otherness'. The splice, on the other hand, implies a friendly merger, a sympathetic co-existence, an integration that dissolves boundaries and distinctions. This latter perspective is reflected in the posthuman view which "configures human being so that it can be seamlessly articulated with intelligent machines. In the posthuman, there are no essential differences or absolute demarcations between bodily existence and computer simulations, cybernetic mechanism and biological organism, robot teleology and human goals" (Hayles 1999, pp. 2–3).

In addition to the cyborg as a symbiotic merger of disparate entities, it also represents the capacity for potentiation. Bruce Clarke's "neocybernetic posthuman is the human metamorphosed by reconnection to the worldly and systemic conditions of its evolutionary possibility" (B. Clarke 2008, p. 196). As "the site of *possible* being" (Gonzalez 1995, p. 267), the concept of cyborg demands exploration of liminal conditions; it constitutes an entity in flux, one that is partial and not whole, an identity steeped in process rather than fixity. In other words, the cyborg body continually re-presents. This aspect of multiplicity and flux in no way negates, undermines, or otherwise deconstructs the human; it simply recognizes and celebrates the symbiotic connections between subject, environment, and what is produced by those processes. Cyborg identity does not take away from our humanity—it enriches it.

Being/becoming a cyborg doesn't undermine our humanness; since the advent of technology we have always been cyborgs in degree. Acknowledging this is only to recognize that humans are not pure, undiluted entities, that we are greater than the assemblage of our constituent body parts (Merleau-Ponty). What makes us human is not just genes but our ingenuity: our ability to grow, to adapt and innovate, to appropriate externalities; it is this capacity that insures our survival, that allows us to endure. Contrary to threatening or even annihilating our humanity, to stifle these attributes would be tantamount to a denial of our humanity.

6.7 Ethical Issues and Considerations

6.7.1 Political Status of the Posthuman Cyborg

What impact do these changes in human identity, both ontologically and epistemologically, have on our political status? Citizenship, and the rights it conferred, was historically enjoyed only by individuals of stature; whiteness, maleness, property ownership and wealth were required attributes for citizenship. (Women, children, and blacks were themselves property of men and as such were not eligible for the protection of rights under the law.) As qualification for citizenship became more inclusive, it was conferred by membership within a state, and then later by membership within the human race. Gray notes that "more and more rights are being linked to a person's status as a human being and not to their membership in a state" (Gray 2002, pp. 22–23). But he sees affiliation to state structures as an inadequate basis for citizenship status: "Since nation-states are contingent, based on belief and history, there is no reason not to imagine a polity of the world. World citizenship makes much more sense than does citizenship in nation-states" (Gray 2002, p. 23).

While eligibility for citizenship has become more expansive as society's collective values evolve—extending from a privileged class of individuals based on gender, race, and class, to subjects of a nation-state, to all members of the human species—eligibility must be revisited again with the posthuman cyborg. Both Gray and James Hughes take the position that posthuman cyborgs should be endowed with citizenship, regardless of their genetic composition—a departure from Annas' view that dignity resides exclusively with humanness (Annas 2000). According to Gray, the determinant for citizenship is the ability for meaningful communication. He writes that "the definition of citizenship is freeing itself from gender-, race-, and class-based criteria and becoming an issue of competent participation in what some philosophers call a discourse community but what most of us would just label a meaningful conversation. The communication need not be speech or writing, as Helen Keller proved, but there must be communication for political participation" (Gray 2002, p. 22). His view is grounded entirely in communication as a political act: "I trace my citizenship to my consent to be governed, in the formulations of John Locke, and that stems from my ability to be part of the polis, the political entity that humans and cyborgs (and in the future who knows what else) share through our ability to communicate together about political issues" (Gray 2002, p. 23).

Gray suggests that the capacity for communication can be ascertained by the same test Turing invented for evaluating whether machines had achieved the equivalent of human intelligence. "The complications of cyborg citizenship call for a cyborg citizen Turing test to determine which entities can participate in our discourse community and which cannot" (Gray 2002, p. 24). He describes the rules of such a cyborgian version: "the best way to judge [intelligence] is by testing whether or not the entity in question could carry on an intelligent conversation with an intelligent human for a serious length of time. If it could, the machine should be considered intelligent, as least as intelligent as many humans. . .An entity must convince a simple majority of 12 other citizens that it can be part of their conversation. This requirement should prevent refusal of citizenship on the basis of racism or other prejudices" (Gray 2002, p. 25). He is careful however, to allow only the inclusion of any entities that can be deemed to possess human levels of intelligence; "the point is not to include pets or fetuses or corporate entities. If such entities deserve rights and protection, they can be granted in other ways than citizenship" (Gray 2002, p. 26). As it has been throughout human history, citizenship for Gray is clearly still meant to be exclusive and absolute.

The problem with Gray's reasoning is its myopic anthropocentrism, the very thing that defeated attempts to develop AI. The Turing test perpetuates exclusivity based on the tyranny of the standard of human intelligence. Is it possible to conceive of types of intelligence that are not comparable to humans? And is it fair to dismiss an entity if it cannot communicate in ways that can be understood by humans? Moreover, since meaning is a social construct, it is doubtful that there could be consensus on what constitutes 'meaningful conversation.' While human-type intelligence should neither be a necessary or sufficient prerequisite for citizenship, even if one does accepts Gray's premise, his view still seems too restrictive. What if Anne Sullivan had not come into Helen Keller's life and discovered how to communicate with her on *her* terms? What if no technology existed that allowed Stephen Hawking to vocalize his scientific insights? Would we really want to conclude that their inability to communicate (or perhaps to put it more precisely, their inability to make themselves understood to other human beings) makes them less eligible for citizenship?

I would argue that there is not a bar of intelligence that, if surpassed, gains one membership in the club of citizenship, and those who don't meet that standard walk away empty-handed. Intelligence can be demonstrated on myriad levels; we have only to observe how living things interact in their environments in ways that are arguably much more 'intelligent' than what humans do. My dogs, for example, can tell whether a stranger has ill will or the intention to harm me well before I am able to discern it. They have a different range of capabilities that I simply do not possess, and they might judge my deficient sensibilities as primitive, if they were so inclined. It seems to be more appropriate to speak of diverse or even graduated intelligence, and thus diverse or graduated citizenship, rather than citizenship as an all-or-nothing proposition. While Gray makes a perfunctory statement that animals should indeed enjoy some limited rights and protections (though he never specifies on what grounds), he does not present any argument as to why they should be *disqualified* from the possibility of citizenship, even to a lesser extent. At the very least a category of second-tier citizens (as opposed to second-class citizens and the pejorative connotations associated with that term) could extend some degree of citizenship—and therefore agency—to other entities with non-human intelligence.

Ironically, Gray observes that "inevitably categories such as Asian, white, black, and Latino collapse under scrutiny, just as today the category is collapsing under cyborgization. Still, for every border trespass there are new mobilizations that futilely attempt to keep identities unsullied and unified" (Gray 2002, p. 192). Yet while he attempts to violate existing borders, the standards he proposes also attempt to preserve an exclusivity of new categories.

Hughes offers a different justification from Gray for the inclusion of posthumans as citizens. Like Gray, he argues that the state of being human should not be the pivotal condition: "humans and animals share almost all their DNA, and most of that small amount of DNA unique to human beings is irrelevant to our specialness...Being furless isn't one of the essential aspects of humanness" (J. Hughes 2004, p. 93). Moreover, he claims that the standard of humanness espoused by the bioluddites is becoming obsolete. For bioluddites, "'humanness' is the green card to citizenship...[but] human/machine cyborgs, xenografts (transplants of animal organs), artificial organs, embryo research and brain alterations... [already] threaten the integrity of the human species itself" (J. Hughes 2004, pp. 78–79). Where he diverges with Gray is in his belief that "citizenship should be based [not on communication, but] on the concept of 'personhood,' of having feelings and consciousness" (J. Hughes 2004, p. 75). Hughes is careful to distinguish the two concepts when he states that "the rights of citizenship are due to persons, not humans, that not all humans are persons, and that humanness is not a necessary condition for personhood" (J. Hughes 2004, p. 83).

While I agree with Hughes that thoughts, feelings, and self-awareness should be the constituting factors for personhood (which would then include posthuman cyborgs and perhaps future AIs and ALs), and that personhood—rather than common biological traits—should endow an entity with citizenship status, I diverge from his postulation of a continuum of rights—not in principle, but in its reductive classifications. The thresholds between categories are not as definitive as he depicts them. While he distinguishes levels of consciousness—from non-sentient to sentient to self-aware to possessing full consciousness—he assigns rights to lower entities based on their status as property and affords no rights at all to entities that are not sentient. In addition, he distinguishes between full and disabled citizens (the latter including children and apes), which implies that if the level and quality of thought and feeling is not equivalent to adult human capacities—what he refers to as 'cognitive equals'—then these entities are subpar. Furthermore, the term 'disabled citizen' that he assigns to children and those with diminished mental capacity connotes a disparaging meaning. Another critical oversight is that the aspect of interdependence between entities at all levels is not considered at all. Despite its limitations, the concept of a continuum is still an appropriate one though, and at least in Hughes model of citizenship, had Helen Keller and Stephen Hawking remained incommunicative to the outside world, they would not have been completely excluded from citizenship as they would have been in Gray's model. But in Hughes' continuum they would still have been classified as disabled citizens.

Hughes' concept of 'cyborg citizenship' may include, in principle, any entities which meet the criterion of personhood: the possession of feelings or consciousness. This criterion opens up possibilities for determining whether a wide spectrum of non-human entities should be entitled to rights. "If the reason why we recognize the rights of persons is because they are intelligent, have feelings, are conscious, and so forth, then the question becomes whether AIs or whales or alien beings share these qualities" (J. Hughes 2004, p. 105). But this begs a fundamental question: by what means can we determine if something has feelings or consciousness? Hughes offers no Turing test by which to evaluate those traits as was proposed in Gray's concept of the citizen subject. At best we can only infer that others have consciousness and feelings through observation and comparison with our own sensibilities. When noted evolutionary biologist Richard Dawkins spoke at the University of Hawai'i a few years ago, he was asked whether he believed animals possessed feelings or consciousness. His answer was surprising in its lack of scientific rationale: if animals displayed emotions like fear or pain in a way that appeared similar to what he experiences, then he would say, yes, they do. I am inclined to believe this as well. Isn't this the same method by which we read others' feelings? And which disposes us to empathy? But they do not need to be *human* feelings. Dog feelings, whale feelings, and elephant feelings are equally valid though they may or may not be as complex or varied or nuanced as human feelings—or perhaps they are, but simply different from ours. Though Hughes recognizes their sentience, his stance falls short in his limited attribution of nothing more complex than the experience of pleasure and pain to animals.

6.7.2 Political Status of the Extraterrestrial Environment

In the context of Space habitation, the discourse on citizenship and rights must necessarily extend beyond the posthuman subject. The terraforming debate, for example, raises some of the unique issues that we will confront in our further exploration of the solar system: if we discover first-genesis life in the form of micro-organisms on Mars—an event that would have profound implications—would they or should they be considered more valuable than our common Earth bacteria? If so, should they be subject to greater protections than the humans who have come to trespass and occupy their indigenous habitats?[9] Gaia and symbiotic theory reveal the interdependence between all living things to each other as well as to their environment. Posthuman theory argues along similar lines that the cyborg is a condition of liminality. Even biologists cannot identify any definitive threshold between living and non-living things. The categorization of 'citizenness,' either in totality or in degree, would appear to run counter to these theories in its arbitrary boundary demarcations.

Everything in Nature—living or not—is, by virtue of its ecological interdependence, vital at some level. Glaciers provide habitats for arctic fauna, but they also elevate the Earth's albedo which helps maintains its climatic homeostasis. To that extent, everything that has evolved has at the very least a right to exist, to be self-actualized, to fulfill its potential or purpose. This is the essence of biocentric equality, a basic principle of the deep ecology philosophy, which acknowledges that all things in the biosphere, both living and non-living, have intrinsic value and an equal right to 'self-realization' (Devall and Sessions 1985). Some may argue that this may take the inclusive position too far to the extreme, but to dismiss all entities that are not comparable in intelligence, feelings, thoughts, self-awareness, consciousness, or communicative ability to human beings is as dangerously absurd as it was when individuals were dismissed from the realm of citizenship on the basis of gender, race, or class. This line of reasoning supports the integration approach of respecting the environment: acknowledging our relationships to it as collaborative in nature, and not treating it as something to be exploited or dominated.

[9] In the film *Aliens*, the overriding directive of the Weyland-Yutani Corporation—who sponsored the mining expeditions to Space—was to keep the alien specimen intact at all costs, even at the expense of the entire human crew (though this was for reasons of commercial and scientific exploitation rather than preservation of the specimen for the sake of itself).

6.7.3 Political Status of Artilects and Other Non-biological Entities

As with all of our technologies, human evolution will be intricately intertwined with robotic evolution. But just what direction it will take is only speculative. Expressing a sentiment that echoes Kurzweil, de Garis writes: "a lot of people will start seeing humanity as a stepping-stone toward a higher form of evolution and will claim it is humanity's destiny to help the artilects get off the planet and into their true environment—namely, the cosmos—perhaps in search of other hyperintelligences" (de Garis 1989, p. 20).

Echoing the sentiments in posthumanist literature, Ramey states that "one need not be troubled by whether the notion of self will become more alienated in an increasingly technologized world populated by strange cyborgs and futuristic androids because there is no faithfully isolated inner self in the first place," and that "an android cannot be truly humanlike until it is regarded as part of a reciprocal relationship in which one's well-being is tied to the maintenance of this Mitsein" (Ramey 2005, pp. 142–143).

Yet another view is that instead of making robots be more like humans, "some human beings might want to modify their own bodies and brains to become artilects themselves" (de Garis 1989, p. 20).

Duffy maps out different possible futures for robotics: "(a) machines will never approach human abilities; (b) robots will inevitably take over the world; (c) people will become robots in the form of cyborgs. A fourth possibility exists. Robots will become a 'race' unto themselves. In fact, they already have" (Duffy 2003, p. 187). These possible scenarios are not incompatible with each other or exclusive, so it may even be possible that all of them are (or will come) true.

How humans and robots co-evolve will depend on their ethical status. Despite the affections that humans have demonstrated for their robotic counterparts, participants in studies did not ascribe the same moral agency to robotic pets as they did to a real animal (Walters et al. 2008). "Robots are artefacts and therefore, in the eyes of many, they have no element of consciousness, which seems to be widely regarded as the dividing line between being deserving of ethical treatment and not" (Levy 2009a, p. 210). But in the context of relations of alterity, the justification for ethical relations is not grounded in the question of consciousness. Coeckelbergh suggests that the discourse shift the focus "from mind to interaction, from intelligent thinking to social-emotional being, from reality to appearance, from right to good, and from theory to experience and imagination" (Coeckelbergh 2009, p. 217).

Trying to ascertain whether robots have consciousness or can attain human-type intelligence is fundamentally irresolvable, since we don't understand enough about either to make any definitive determination. But what we can be certain about is how they *affect* us. "What robots do to us, depends on how they appear to us, not on what is 'really' in their mind. For instance, existing robots are not considered sentient or in possession of feelings; nevertheless, when humans interact with some types of robots they may act and talk *as if* the robot has sensations and feelings. In a similar way, humans tend to attribute thoughts and beliefs to robots" (Coeckelbergh 2009, p. 219). One can argue that the intelligence of robots is an illusion or is 'faked,' but that is irrelevant as it is the *relations* that matter. In relationships, two entities offer a different perspective of the world to each other (Ramey 2005); when two beings can engage in this manner, this is what Lee refers to as a 'subject-of-a-life' (Lee 2006, p. 426). Robots can engage us intellectually and emotionally, they can allow us to see the world anew, and they can help us do what we could not do alone—is this not sufficient basis for advocating that ethical treatment be extended to robots and artilects? If we do not acknowledge the robot, artilect, or android as 'significant other,' we do not only degrade robots but we also debase ourselves as humans. To treat them merely as dispensable objects is to "dehumanize the self and ignore the inherently ethical relationship among humans in which these androids enter" (Ramey 2005, p. 137). Our interaction with them necessarily has ethical consequences, and through this 'Mitsein' it changes our sense of who we are and our being in the world.

Ascribing agency to robots is often accompanied by the fear that in doing so, robots may take over the world or ultimately cause us harm when they become more intelligent than us, a sentiment that is expressed in films such as *The Terminator*. But this concern stems from the misguided ontological view of robots, coupled with the Darwinian mindset that natural selection will favor artificial beings over organic ones, both of which are rendered baseless when we look at the issue as a relational one. What robots do to us, we also do to them. We make them intelligent and social beings though our exchanges, not through their software. Whether machines become sentient or not, robots do not threaten our humanity—they can reinforce and enrich it. They offer us the opportunity to make ourselves better than what we are now, if we choose to do the same for them.

6.8 Summary

The implications of migrating to Space for human evolution are extensive and have little to do with strictly conventional biological processes. If change was measured exclusively by the content of our DNA as we do with all other forms of life, questions about what makes us human (in this case, the species of *Homo sapiens*) and what is the threshold of speciation would be easy and straightforward to answer using scientific methods of genetic testing.

But our genome is only one facet of what makes us human. The aggregate qualities of 'humanness' encompass much more than the standard attributes of *Homo sapiens*. The ontology of human being is a complex amalgam of not only our material/physical instantiation, but also our capacities and our relationships, entailing a political and ethical dimension. Being human is better defined as a process than a product. Whether we choose to fuse ourselves with our technologies, migrate to distant not-Earth environments, or deliberately modify our genetic composition, we are always 'becoming human.'

References

Annas, G. J. (2000). Man on the moon, immortality, and other millennial myths: The prospects and perils of human genetic engineering. *The Emory Law Journal, 49*, 753.

Capra, F. (1996). *The web of life: A new scientific understanding of living systems*. New York, NY: Anchor Books.

Clark, A. (2003). *Natural-born cyborgs: Minds, technologies, and the future of human intelligence*. Oxford, England: Oxford University Press.

Clarke, J. (2005). A sensorial act of replication. In M. Smith (Ed.), *Stelarc: The monograph* (pp. 193–212). Cambridge, MA: The MIT Press.

Clarke, B. (2008). *Posthuman metamorphosis: Narrative and systems*. New York, NY: Fordham University Press.

Coeckelbergh, M. (2009). Personal robots, appearance, and human good: A methodological reflection on roboethics. *International Journal of Social Robotics, 1*(3), 217–221.

Crosby, A. (1986). Life (with all of its problems) in space. In B. Finney & E. Jones (Eds.), *Interstellar migration and the human experience* (pp. 210–219). Berkeley, CA: University of California Press.

Dator, J. (2012). *Social foundations of human space exploration*. New York, NY: Springer.

de Garis, H. (1989). What if AI succeeds? The rise of the twenty-first century artilect. *AI Magazine, 10*(2), 17.

DeLanda, M. (2006). *A new philosophy of society: Assemblage theory and social complexity*. London, England: Continuum.

Devall, B., & Sessions, G. (1985). *Deep ecology: Living as if nature matters*. Salt Lake City, UT: Gibbs Smith.

Duffy, B. R. (2003). Anthropomorphism and the social robot. *Robotics and Autonomous Systems, 42*(3), 177–190.

Finney, B. (1986). The exploring animal. In B. Finney & E. Jones (Eds.), *Interstellar migration and the human experience* (pp. 15–25). Berkeley, CA: University of California Press.

Finney, B., & Jones, E. (Eds.). (1986). *Interstellar migration and the human experience*. Berkeley, CA: University of California Press.

Gonzalez, J. (1995). Envisioning cyborg bodies. In C. H. Gray (Ed.), *The cyborg handbook* (pp. 267–279). New York, NY: Routledge.

Gray, C. (Ed.). (1995). *The cyborg handbook* (1st ed.). New York, NY: Routledge.

Gray, C. H. (2002). *Cyborg citizen: Politics in the posthuman age*. New York, NY: Routledge.

Halberstam, J., & Livingston, I. (Eds.). (1995). *Posthuman bodies*. Indianapolis, IN: Indiana University Press.

Haraway, D. J. (1991). *Simians, cyborgs, and women: The reinvention of nature* (1st ed.). New York, NY: Routledge.

Hart, M. (1986). Interstellar migration, the biological revolution, and the future of the galaxy. In B. Finney & E. Jones (Eds.), *Interstellar migration and the human experience* (pp. 278–292). Berkeley, CA: University of California Press.

Hayles, N. K. (1999). *How we became posthuman: Virtual bodies in cybernetics, literature, and informatics* (1st ed.). Chicago, IL: University of Chicago Press.

Hughes, J. (2004). *Citizen cyborg: Why democratic societies must respond to the redesigned human of the future*. New York, NY: Basic Books.

Jain, S. (2006). The prosthetic imagination. In M. Smith & J. Morra (Eds.), *The prosthetic impulse: From a posthuman present to a biocultural future*. Cambridge, MA: MIT Press.

Kelly, K. (1994). *Out of control: The new biology of machines, social systems, and the economic world*. New York, NY: Basic Books.

Kelly, K. (2010). *What technology wants*. New York, NY: Viking Adult.

Lee, B. (2006). Empathy, androids and "authentic experience". *Connection Science, 18*(4), 419–428.

Levy, D. (2009a). The ethical treatment of artificially conscious robots. *International Journal of Social Robotics, 1*(3), 209–216.
Malabou, C. (2008). *What should we do with our brain?* New York, NY: Fordham University Press.
Margulis, L., & Sagan, D. (1997b). *Microcosmos: Four billion years of microbial evolution* (1st ed.). Berkeley, CA: University of California Press.
Pepperell, R. (2003). *The post-human condition.* Portland, OR: Intellect Books.
Ramey, C. H. (2005). For the sake of others: The personal ethics of human-android interaction. In *Proceedings of the CogSci 2005 Workshop: Toward Social Mechanisms of Android Science* (pp. 137–148).
Smith, M. (Ed.). (2005). *Stelarc: The monograph.* Cambridge, MA: MIT Press.
Stock, G. (2002). *Redesigning humans: Choosing our genes, changing our future.* New York, NY: Houghton Mifflin Books.
Stone, A. R. (1996). *The war of desire and technology at the close of the mechanical age.* Cambridge, MA: MIT Press.
Tanner, N. M. (1986). Interstellar migrations: The beginnings of a familiar process in a new context. In B. Finney & E. Jones (Eds.), *Interstellar migration and the human experience* (pp. 220–233). Berkeley, CA: University of California Press.
Valentine, J. (1986). The origins of evolutionary novelty and galactic colonization. In B. Finney & E. Jones (Eds.), *Interstellar migration and the human experience* (pp. 266–277). Berkeley, CA: University of California Press.
Walters, M. L., Syrdal, D. S., Dautenhahn, K., Te Boekhorst, R., & Koay, K. L. (2008). Avoiding the uncanny valley: Robot appearance, personality and consistency of behavior in an attention-seeking home scenario for a robot companion. *Autonomous Robots, 24*(2), 159–178.

Chapter 7
Conclusion

7.1 Summary of Outcome

This dissertation is comprised of three major sections: Space exploration objectives, the technological means to achieve those ends, and the implications of our strategies on human evolution. The first section focuses on the expanding exploration goals in Space, beginning with humans' first missions to the Moon, then to the extended length of stay on the Space stations, and finally the initiative to send the first humans to Mars. The timeframe expanded from weeks and months to years, mandating reassessments of design criteria. Only survival strategies for short-term adaptation were relevant to the brief missions in the earliest phase of human Space exploration. The longer stays on the ISS and the proposed excursions to Mars began to raise issues of habitability for mid-term adaptation. But an even larger outlook decades or generations in the future that espouses evolutional factors for long-term adaptation has not yet warranted any attention.

The second section evaluates the various technological means by which we can best achieve those adaptation ends. A preliminary discussion on how technology mediates the relationship between humans and their environment served as the basis for the development of my position. This was followed by presentation of specific types of technologies that would be viable for creating various interfaces with extraterrestrial environments. Four archetypal approaches based on intentionality were identified and compared in the exploratory forecasting component of the argument. Shifting to the normative forecasting component I presented an argument in favor of integration, and expounded on various theories in evolution, environment, cooperation, and cosmology to support it.

The third section examined the ontological implications of the technological means that we employ. Design proposals are often implemented because they are practical in the moment, or useful to solve a very specific problem, but without consideration of the wider range of implications of those applications. The precautionary principle advises that all foreseeable impacts are evaluated before embarking on any course of action. This section studies how the technologies we

E.S. Lockard, *Human Migration to Space*, Springer Theses, 185
DOI 10.1007/978-3-319-05930-3_7, © Springer International Publishing Switzerland 2014

adopt, and the manner in which they are implemented, will affect human evolution in Space. Both technological intervention as well as the radically different environmental conditions of the Moon or Mars will precipitate profound evolutionary consequences, which then sets the stage for discussion of what we will become and how that transformation affects political and ethical status.

At present human factors research addresses issues of habitability. An initial assessment of habitability criteria revealed the assumption that habitability entails only the provision of creature comforts, pleasurable activities (as a reprieve from work activities), and physiological and psychological well-being. These are certainly important aspects of habitability but by no means are they sufficient. Although the incorporation of habitability standards as part of human factors design criteria is moving in the right direction, it is intended only to achieve mid-term adaptation—a span of only a few years (several at best), though certainly not for an entire lifetime, and less so for subsequent generations.

A thoughtful investigation of the evolutional factors entailed by long-term adaption—the needs over a lifetime as well as for future generations—was necessary. This distinction of levels of adaptation corresponds with Maslow's pyramid illustrating the hierarchy of human needs (Fig. 7.1). Grounded in the phenomenological research, and informed in large part by my training as an architect, I posited that an experiential component was crucial to long-term adaptation. In order to feel fulfilled, to create meaning and purpose in one's daily life, and to develop a sense of achievement, humans need to be engaged with the resistances of their environment. We rely on processes of active negotiation, not comfortable passivity, to become familiar with the places we inhabit and dwell in, and therefore the need for engagement in order to insure long-term adaptation should not be underestimated.

7.2 The Evolution of *Homo Caelestis*

Humans will not evolve in Space similarly, even if using exactly the same technologies. Not only the differing environmental conditions, but also the associations with the planets, will influence the directions for the possible futures of humans who live there.

7.2.1 Homo Caelestis Martius

The celestial human of Mars will be represented by the first migrants to Space. As the seminal community to live beyond Earth, they will be the original prototype and precedent-setter for subsequent settlements. The astronaut pioneers will enjoy a heroic celebrity status, much like Neil Armstrong and Buzz Aldrin enjoyed as the first humans to walk on the Moon. Due to the novelty of living in Space and the high risks involved in these first ventures, the crew will be subject to rapt attention; every

Fig. 7.1 Maslow's hierarchy of needs

event will become public knowledge. This elevated level of interest and scrutiny will likely be stressful for the crew members, but on the other hand may also help attenuate feelings of isolation.

Given the distance of Earth from Mars, return to the homeplanet will be rare, or even unlikely. This sense of permanence will influence how the crew will assimilate to their adopted planet. They will be more inclined to think of Mars as their new home and to engage with their environment more deeply. The fact that the conditions are relatively similar to Earth in comparison with the other possible habitation venues (partial gravity, an almost aligned diurnal cycle, varied geological features such as mountains and canyons, and the presence of life-essential water), will help the crew feel more connected to the planet.

7.2.2 Homo Caelestis Luna

The celestial human of the Moon will be constituted by the second generation of humans in Space. Establishing a presence on another planetary body will hold some excitement for the public, but because this will be the second wave of migration, as well as to a place where humans have already traveled, the fervor that accompanied the ground-breaking events of Mars will be somewhat diminished.

In addition, the concomitant risks will be lower, having benefited from the lessons learned on the Martian outposts, and the shorter distance makes commuting back to Earth more probable. On the other hand, the lunar conditions will be more challenging than on Mars: the radically different diurnal cycles make entrainment prohibitive, there is much less gravity—which means physical actions will be executed more slowly—and there is even less atmosphere than on Mars, so there will be no weather conditions whatsoever.

Other aspects which affect the Moon's status is the fact that it is a small satellite rather than a full-fledged planet, and perhaps more importantly, it has no history of life—nor is it capable of sustaining any. Given all these factors, the Moon may be perceived as a second-class suburb to Earth.

7.2.3 Homo Caelestis Tracta Externus

The celestial human of outer Space—in other words, without a planet—is a third possibility. Dwelling in LEO however, has a more demanding set of challenges. On a physical level, there is only absence: no gravity and therefore no resistance, no resources, no other living or even non-living forms—just open space. Cycles which are essential to establishing life's circadian rhythms—diurnal cycles, monthly cycles, seasonal cycles, yearly cycles—will be non-existent. Patterns of time will need to be artificially induced by the built habitat.

Furthermore, without terra firma, there will always be a sense of placelessness, of being nowhere. Having a sense of location is fundamental to developing the phenomenal condition of home. The habitat cannot establish an interface with the environment, and thus comprises the entire sensible realm for its occupants. Sans location, any habitat in LEO will be perceived as a station rather than an outpost, as temporary rather than permanent, and being en route as opposed to being a destination.

All of these issues can be countered only if something resembling substantial terra firma that can provide a referential exteriority is created and its location can be mapped or coordinated. This is a tall order that would require producing something akin to an artificial planetoid or satellite as an environmental context for the habitat.

7.3 Possible Directions for Further Research and Development

Any good thesis should raise the discourse to another level, which then in turn frames a new set of questions. This project opens the doors for several different directions that could be researched further as an appendix to the work I have begun. I have purposely avoided attempting to describe what the human of the future will or should resemble, as my focus was on prescribing the kind of relationships with the extraterrestrial environment that should be cultivated in order to achieve adaptation; hence it was more process-oriented than product-oriented. However, an exploratory approach could similarly be pursued to forecast what humans might become, or a normative forecasting approach to define what humans in Space should become, specifying the technological means to achieve the preferred end.

In the case of this dissertation, possibilities for achieving integration have been suggested, but should such an approach be desired, more detailed steps are necessary. First, new technologies and advances in existing ones constantly emerge, which require frequent reassessment. Other technological options may present themselves besides the ones I have enumerated in this document. Second, in addition to the technological strategies I expound upon, political, social, and economic strategies—as well as potential obstacles that may be encountered and how to overcome them—require more consideration than I have given them in this project. Third, parameters will change for future phases of habitation in Space, thus what promotes adaptation for one generation or locale, will be different for another generation and locale. There is still much to be explored in the research prior to further exploration of Space.

7.4 Reflections

I did not initially set out to write a dissertation on long-term adaptation in Space; in fact, the idea had never crossed my mind until much later on in my research. Beginning only with a critical stance of the myopic engineering perspective towards Space architecture, I originally embarked on the subject of human habitat design in Space from the perspective of improving functionality and the aesthetics of the living environment. However, I quickly discovered that habitat design criteria and other human factors considerations had already been extensively studied and written about. But as I continued to scan the literature, the new direction on adaptation was formulated as a synthesis of pervasive deficiencies pertaining to human experience, aesthetics, and a futures perspective.

Rather than pursuing a dialectical methodology, my approach to my theoretical development is decidedly more inclusive. The underlying agenda in both my writing and in my design work has always been unapologetically one of reconciliation of ostensibly disparate themes, which is why I embraced symbiosis and cooperation theories long before I had developed the actual premise for my dissertation. Avoiding the divisiveness and contrariness that is so often the norm in scholarly discourse, I sought to distill commonalities and compatibilities upon which to strengthen my position. This disposition should be evident in the conclusions I draw in my dissertation.

I expect to witness the first human habitation in Space within my lifetime, and when the time comes I will anticipate it with both feelings of exultation and trepidation. The regressive state of current social relations, the capriciousness of our capacity for compassion, the human-induced maladies that plague the environment, and the destructive tendencies of humankind in general, sometimes leave me pessimistic about our chances to do something good in Space. On the other hand, there is enormous opportunity in this next phase of human exploration to elevate the human condition, to bring out something transcendent in our species, and to continue to discover the magnificence of the cosmos. I hope that this dissertation will be a small contribution to that noble effort.

Acknowledgements

I'd like to express my appreciation to my committee members for the mix of support, scholarly guidance, and 'tough love' they gave me in completing this dissertation:

Jim Dator, who introduced me to a new world of futures studies, and encouraged me by example to embrace the absurd in my academic pursuits.

Debbie Halbert, whose uncompromising standards pushed me to improve my work beyond what I thought I was capable of.

Kim Binsted, who embodies the pioneering spirit and grounded my efforts with a Space scientist's perspective.

David Swift, whose ebullient enthusiasm for all things Space fueled my interest, and who otherwise provided me with much needed moral support throughout the process.

Mike Shapiro, without whom I would not have persevered in completing this project. He patiently guided me along the way, and gave me hope at a time when I was ready to abandon my efforts.

And to the other people who played an important role during my journey:

William Kramer, who took a similar journey into Space with me. He instilled the wonders of biology into this non-scientist, as well as proofread my final draft when I could not read it one more time.

Gayle Lanthier, the friend and confidante I am blessed to have, and who is always there for me.

Arthur Lockard, who not only provided an ear for my inchoate ideas throughout the years but also generally gave me sound advice on all the important matters in my life. He was also the person who, no matter how discouraged I became, could always lift my spirits.

E.S. Lockard, *Human Migration to Space*, Springer Theses,
DOI 10.1007/978-3-319-05930-3,

Bibliography

Anderson, W. T. (1987). *To govern evolution: Further adventures of the political animal.* Orlando, FL: Harcourt-Brace Jovanovich Publishers.

Annas, G. J. (2000). Man on the moon, immortality, and other millennial myths: The prospects and perils of human genetic engineering. *The Emory Law Journal, 49*, 753.

Arendt, H. (1998). *The human condition.* Chicago, IL: University Of Chicago Press.

Arkin, R. C., Fujita, M., Takagi, T., & Hasegawa, R. (2003). An ethological and emotional basis for human–robot interaction. *Robotics and Autonomous Systems, 42*(3–4), 191–201.

Armstrong, R. (Ed.). (2000). *Space architecture.* London, England: Wiley-Academy Press.

Arthur, W. B. (2009). *The nature of technology: What it is and how it evolves.* New York, NY: Free Press.

Axelrod, R. (1987). *The evolution of cooperation.* Orlando, FL: Harcourt-Brace Jovanovich Publishers.

Axelrod, R., & Keohane, R. (1986). Achieving cooperation under anarchy: Strategies and institutions. In K. Oye (Ed.), *Cooperation under anarchy* (pp. 226–254). Princeton, NJ: Princeton University Press.

Ayre, M., Menon, C., & Ellery, A. (2006). Biomimetics – A new approach for space system design. *ESA Bulletin, 125*, 20–26. Retrieved from www.adsabs.harvard.edu/abs/2006ESABu. 125

Bachelard, G. (1994). *The poetics of space: The classic look at how we experience intimate places.* Boston, MA: Beacon.

Balkin, J., & Noveck, B. (2006). *The state of play: Law, games, and virtual worlds.* New York, NY: NYU Press.

Bannova, O., Landschulz, A., & Smith, I. F. C. (2005). Autonomous architecture: Summit station in Greenland design proposal as a test-bed for future planetary exploration. *SAE Technical Paper No. 2005-01-2909.* Warrendale, PA: SAE International.

Basalla, G. (1989). *The evolution of technology.* Cambridge, MA: Cambridge University Press.

Bates, D., & Plog, F. (1991). *Human adaptive strategies.* New York, NY: McGraw-Hill.

Bateson, G. (1963). The role of somatic change in evolution. *Evolution, 17*(4), 529.

Bell, S., & Morris, L. (Eds.) (2009). *Living in space: Cultural and social dynamics, opportunities, and challenges in permanent space habitats.* Aerospace Technology Working Group.

Benaroya, H. (2010). *Lunar settlements.* Boca Raton: CRC Press.

Bennett, J., Shostak, S., & Jakosky, B. (2002). *Life in the universe.* San Francisco, CA: Pearson/ Addison Wesley.

Benyus, J. M. (2002). *Biomimicry: Innovation inspired by nature.* New York, NY: Harper Collins.

Binsted, K., Kobrick, R. L., O'Griofa, M., Bishop, S., & Lapierre, J. (2010). Human factors research as part of a Mars exploration analogue mission on Devon Island. *Planetary and Space Science, 58*(7–8), 994–1006.

E.S. Lockard, *Human Migration to Space*, Springer Theses,
DOI 10.1007/978-3-319-05930-3, © Springer International Publishing Switzerland 2014

Biocca, F. (1997). The cyborg's dilemma: Progressive embodiment in virtual environments. *Journal of Computer-Mediated Communication, 3*(2).
Bird, J., Curtis, B., Mash, M., Putnam, T., Robertson, G., & Tickner, L. (1994). *Travellers' tales: Narratives of home and displacement*. London: Routledge.
Bishop, S. L. (2006). Psychological and psychosocial health and well-being at pole station. In C. Cockell (Ed.), *Project Boreas: A station for the Martian geographic North Pole* (pp. 160–171). Liverpool, England: British Interplanetary Society.
Bishop, S. L., Kobrick, R., Battler, M., & Binsted, K. (2010). FMARS 2007: Stress and coping in an arctic Mars simulation. *Acta Astronautica, 66*(9), 1353–1367.
Blair, S. M. (1991). The Antarctic experience. In A. Harrison, Y. Clearwater, & C. McKay (Eds.), *From Antarctica to outer space* (pp. 57–64). New York, NY: Springer-Verlag.
Bluth, B. J. (1983). Sociology and space development. In T. Cheston (Ed.), *Space social science*. NASA/JSC Software, Robotics and Simulation Division. Retrieved from http://er.jsc.nasa.gov/seh/sociology.html
Boellstorff, T. (2008). *Coming of age in second life: An anthropologist explores the virtually human*. Princeton, NJ: Princeton University Press.
Bowles, S., & Gintis, H. (2003). Origins of human cooperation. In P. Hammerstein (Ed.), *Genetic and cultural evolution of cooperation* (pp. 429–444). Cambridge, MA: MIT.
Bradbury, R. (1997). *The Martian chronicles*. New York City: Harper Collins.
Breazeal, C. (2001). Affective interaction between humans and robots. In J. Kelemen & P. Sosík (Eds.), *Advances in artificial life* (pp. 582–591). Berlin Heidelberg: Springer.
Breazeal, C. (2003). Toward sociable robots. *Robotics and Autonomous Systems, 42*(3–4), 167–175.
Breazeal, C., & Scassellati, B. (2002). Robots that imitate humans. *Trends in Cognitive Sciences, 6*(11), 481–487.
Brin, D. (2006). Singularities and nightmares: Extremes of optimism and pessimism about the human future. Retrieved from https://lifeboat.com/ex/singularities.and.nightmares
Cagle, Y. (2009). *Space adaptation syndrome*. Presentation given at NASA University of Hawaii Astrobiology Institute, Honolulu.
Capra, F. (1996). *The web of life: A new scientific understanding of living systems*. New York, NY: Anchor Books.
Carrere, S., Evans, G. W., & Stokols, D. (1991). Winter-over stress: Physiological and psychological adaptation to an Antarctic isolated and confined environment. In A. Harrison, Y. Clearwater, & C. McKay (Eds.), *From Antarctica to outer space* (pp. 229–237). New York, NY: Springer-Verlag.
Casey, E. S. (1996). How to get from space to place in a fairly short stretch of time: phenomenological prolegomena. In S. Feld & K. H. Basso (Eds.), *Senses of place* (pp. 13–51). Santa Fe, NM: School of American Research.
Chaisson, E. J. (2001). *Cosmic evolution: The rise of complexity in nature*. Cambridge, MA: Harvard University Press.
Chaisson, E. J. (2005). *Epic of evolution: Seven ages of the cosmos*. New York, NY: Columbia University Press.
Clark, A. (2003). *Natural-born cyborgs: Minds, technologies, and the future of human intelligence*. Oxford, England: Oxford University Press.
Clarke, B. (2008). *Posthuman metamorphosis: Narrative and systems*. New York, NY: Fordham University Press.
Clarke, J. (2005). A sensorial act of replication. In M. Smith (Ed.), *Stelarc: The Monograph* (pp. 193–212). Cambridge, MA: MIT.
Clynes, M., & Kline, N. (1960). Cyborgs and space. In C. H. Gray (Ed.), *The cyborg handbook* (pp. 29–33). New York, NY: Routledge.
Coeckelbergh, M. (2009). Personal robots, appearance, and human good: A methodological reflection on roboethics. *International Journal of Social Robotics, 1*(3), 217–221.

Coeckelbergh, M. (2011). Humans, animals, and robots: A phenomenological approach to human-robot relations. *International Journal of Social Robotics, 3*(2), 197–204.
Cohen, M. (1998). Space Habitat Design Integration Issues. *SAE Technical Paper No. 981800*. Warrendale, PA: SAE International. Retrieved from http://papers.sae.org/981800/
Cronin, B. (1999). *Community under anarchy: Transnational Identity and the evolution of cooperation*. New York, NY: Columbia University Press.
Crosby, A. (1986). Life (with all of its problems) in space. In B. Finney & E. Jones (Eds.), *Interstellar migration and the human experience* (pp. 210–219). Berkeley/Los Angeles, CA: University of California Press.
Csikszentmihalyi, M., & Halton, E. (1981). *The meaning of things: Domestic symbols and the self*. Cambridge, England: Cambridge University Press.
Danius, S. (2002). *The senses of modernism: Technology, perception, and aesthetics*. Ithaca, NY: Cornell University Press.
Darwin, C. (2004). *The descent of man*. London, England: Penguin Classic Books.
Dator, J. (2012). *Social foundations of human space exploration*. New York, NY: Springer.
Dawkins, R. (1996). *The blind watchmaker: Why the evidence of evolution reveals a universe without design*. New York, NY: W. W. Norton & Company.
Dawkins, R. (2006). *The selfish gene: 30th anniversary edition*. Oxford, England: Oxford University Press.
de Certeau, M. (1984). *The practice of everyday life*. Berkeley: University of California Press.
de Garis, H. (1989). What if AI succeeds? The rise of the twenty-first century artilect. *AI Magazine, 10*(2), 17.
de Garis, H. (1990). The 21st century artilect: Moral dilemmas concerning the ultra intelligent machine. *Revue Internationale de Philosophie, 44*(172), 131–138.
DeLanda, M. (2006). *A new philosophy of society: Assemblage theory and social complexity*. London: Continuum.
Demiris, J., & Hayes, G. (July 1996). Imitative learning mechanisms in robots and humans. In Proceedings of the 5th European Workshop on Learning Robots, Bari, Italy.
Dennett, D. C. (1995). *Darwin's dangerous idea: Evolution and the meanings of life*. New York, NY: Simon & Schuster.
Devall, B., & Sessions, G. (1985). *Deep ecology: Living as if nature matters*. Salt Lake City, UT: Gibbs Smith.
Diamond, J. (2005). *Collapse: How societies choose to fail or succeed*. New York, NY: Penguin Books.
Dibbell, J. (1998). *My tiny life: Crime and passion in a virtual world*. New York, NY: Holt Paperbacks.
Dimendberg, E. (1995). The will to motorization: Cinema, highways, and modernity. *October* 73:91–137.
Dixon, J. B., & Cassidy, E. (1998). *Virtual futures: Cyberotics, technology and posthuman pragmatism*. New York, NY/London, England: Routledge.
Dodge, M., & Kitchin, R. (2001). Geographies of cyberspace. In *Mapping cyberspace* (pp. 52–64). London: Routledge.
Dollens, D. (2001). *D2A: Digital to analog architecture*. Santa Fe, NM: Lumen Books.
Dollens, D. (2003). *Genetic architectures/Arquitecturas genéticas*. Santa Fe, NM: Lumen Books.
Dollens, D. (2005). *Digital botanic architecture*. Santa Fe, NM: Lumen Books.
Dovidio, J., Gaertner, S., & Esses, V. (2008). Cooperation, common identity, and intergroup contact. In B. Sullivan, M. Snyder, & J. Sullivan (Eds.), *Cooperation: The political psychology of effective human interaction* (pp. 143–160). Hoboken, NJ: Blackwell.
Dudley-Rowley, M., & Bishop, S. L. (October 2002). Extended mission systems integration standards for the human-environment and human-human interfaces. In *Presentation for the AIAA Space Architecture Symposium in Houston, TX*. Reston, VA: American Institute for Aeronautics & Astronautics.

Dudley-Rowley, M., Nolan, P., Bishop, S., Farry, K., & Gangale, T. (2001). Ten missions, two studies: Crew composition, time, and subjective experience in Mars-analog expeditions. In *Proceedings of the Third International Convention of the Mars Society in San Diego, CA*.

Dudley-Rowley, M., Okushi, J., Gangale, T., Flores, P., & Diaz, E. (September 2003). Design implications of latent challenges to the long-duration space mission. In *Proceedings of the AIAA meeting in Long Beach, CA* (Vol. 24). Reston, VA: American Institute for Aeronautics & Astronautics.

Dudley-Rowley, M., Whitney, S., Bishop, S., Caldwell, B., Nolan, P., & Gangale, T. (2002b). Crew size, composition, and time: Implications for habitat and workplace design in extreme environments. *American Institute for Aeronautics and Astronautics, 6111*, 13.

Duffy, B. R. (2003). Anthropomorphism and the social robot. *Robotics and Autonomous Systems, 42*(3), 177–190.

Dyson, G. B. (1998). *Darwin among the machines: The evolution of global intelligence*. Cambridge, MA: Perseus Books.

Edwards, A. R., & Orr, D. W. (2005). *The sustainability revolution: Portrait of a paradigm shift*. British Columbia, Canada: New Society Publishers.

Elton, M. (1997). Robots and rights: The ethical demands of artificial agents. *Ends and Means, 1* (2), 19–23.

Entrikin, J. N. (1991). The betweenness of place. In *The betweenness of place: Towards a geography of modernity* (pp. 6–28). Baltimore, Maryland: Johns Hopkins University Press.

Evans, G., Stokols, D., & Carrere, S. (1987). Human adaptation to isolated and confined environments: Preliminary findings of a seven month Antarctic winter-over human factors study (Final Technical Report).

Fehr, E., & Henrich, J. (2003). Is strong reciprocity a maladaptation? On the evolutionary foundations of human altruism. In P. Hammerstein (Ed.), *Genetic and cultural evolution of cooperation* (pp. 55–82). Cambridge, MA: MIT.

Finney, B., & Jones, E. (Eds.). (1986). *Interstellar migration and the human experience*. Berkeley, Los Angeles, CA: University of California Press.

Finney, B. (1986a). The exploring animal. In B. Finney & E. Jones (Eds.), *Interstellar migration and the human experience* (pp. 15–25). Berkeley, Los Angeles, CA: University of California Press.

Finney, B. (1986b). The prince and the eunuch. In B. Finney & E. Jones (Eds.), *Interstellar Migration and the human experience* (pp. 196–209). Berkeley/Los Angeles, CA: University of California Press.

Finney, B. (1988). Will space change humanity? *Fundamentals of Space Life Sciences, 2*, 247–255.

Fogg, M. J. (1998). Terraforming Mars: A review of current research. *Advances in Space Research, 22*(3), 415–420.

Fogg, M. J. (2000). The ethical dimensions of space settlement. *Space Policy, 16*(3), 205–211.

Fukuyama, F. (2002). *Our posthuman future: Consequences of the biotechnology revolution*. New York, NY: Farrar, Straus & Giroux.

Fuller, R. B. (1971). *Operating manual for spaceship earth*. New York, NY: Penguin Group.

Gaggioli, A., Vettorello, M., & Riva, G. (2003). From cyborgs to cyberbodies: The evolution of the concept of techno-body in modern medicine. *Psychology Journal, 1*(2), 75–86.

Gallagher, W. (2001). *The power of place*. Cambridge, MA: Harvard University Press.

Gangale, T., & Dudley-Rowley, M. (2005). Issues and options for a Martian calendar. *Planetary and Space Science, 53*(14), 1483–1495.

Gardner, J. N. (2003). *Biocosm: The new scientific theory of evolution*. Maui, HI: Inner Ocean.

Gärling, T. E., & Evans, G. (1991). *Environment, cognition, and action: An integrated approach*. Oxford, England: Oxford University Press.

Garreau, J. (2005). *Radical evolution*. New York, NY: Broadway Books.

Gjerde, K. (2000). Terraforming: The modern alchemy—or tomorrow's engineering technology? Retrieved from the *Terraforming Information Pages* at http://www.users.globalnet.co.uk/~mfogg/

Glassman, M. (2000). Mutual aid theory and human development: Sociability as primary. *Journal for the Theory of Social Behaviour, 30*(4), 391–412.

Glenn, J. C., & Gordon, T. J. (2003). *Futures research methodology*. American Council for the United Nations University, the Millennium Project.

Gonzalez, J. (1995). Envisioning cyborg bodies. In C. H. Gray (Ed.), *The cyborg handbook* (pp. 267–279). New York, NY: Routledge.

Goodall, J. (2005). The will to evolve. In M. Smith (Ed.), *Stelarc: The monograph* (pp. 1–31). Cambridge, MA: MIT.

Goonatilake, S. (1999). *Merged evolution: Long-term complications of biotechnology and information technology*. New York, NY: Routledge.

Gray, C. (Ed.). (1995). *The cyborg handbook* (1st ed.). New York, NY: Routledge.

Gray, C. H. (2002). *Cyborg citizen: Politics in the posthuman age*. New York, NY: Routledge.

Gushin, V. I., & Dudley-Rowley, M. (1999). On our best behavior: Optimization of group functioning on the early Mars missions. *The Founding Convention of the Mars Society, Boulder, Colorado* (pp. 717–721).

Halberstam, J., & Livingston, I. (Eds.). (1995). *Posthuman bodies*. Indianapolis, IN: Indiana University Press.

Hall, T. W. (1994). *The architecture of artificial-gravity environments for long-duration space habitation*. Doctoral dissertation, University of Michigan.

Hall, T. W. (2004). *Architectural design to promote human adaptation to artificial gravity*. NASA White paper.

Hammerstein, P. (Ed.). (2003). *Genetic and cultural evolution of cooperation*. Cambridge, MA: MIT.

Haraway, D. J. (1991). *Simians, cyborgs, and women: The reinvention of nature* (1st ed.). New York, NY: Routledge.

Harries, K. (1998). *The ethical function of architecture*. Cambridge, MA: MIT.

Harrison, A. A. (2002). *Spacefaring: The human dimension*. Berkeley: University of California Press.

Harrison, A. A. (2010). Humanizing outer space: architecture, habitability, and behavioral health. *Acta Astronautica, 66*(5–6), 890–896.

Harrison, A., Clearwater, Y., & McKay, C. (Eds.). (1991). *From Antarctica to outer space: Life in isolation and confinement*. New York, NY: Springer-Verlag.

Hart, M. (1986). Interstellar migration, the biological revolution, and the future of the galaxy. In B. Finney & E. Jones (Eds.), *Interstellar migration and the human experience* (pp. 278–292). Berkeley/Los Angeles, CA: University of California Press.

Hawken, P., Lovins, A., & Lovins, L. (1999). *Natural capitalism: Creating the next industrial revolution*. Boston, MA: Little, Brown & Co.

Hawley, A. H. (1950). *Human ecology: A theory of community structure*. New York, NY: The Ronald Press Company.

Hayles, N. K. (1999). *How we became posthuman: Virtual bodies in cybernetics, literature, and informatics* (1st ed.). Chicago, IL: University of Chicago Press.

Haynes, R., & McKay, C. (1992). The implantation of life on Mars: Feasibility and motivation. *Advances in Space Research, 12*(4), 133–140.

Heidegger, M. (2008). *Basic writings*. New York, NY: Harper Perennial Modern Classics.

Heim, M. (1998). *Virtual realism*. Oxford, England: Oxford University Press.

Henrich, J., Bowles, S., Boyd, R., Hopfensitz, A., Richerson, P., Sigmund, K., Smith, E., Weissing, F., & Young, H. P. (2003). Group report: The cultural and genetic evolution of human cooperation. In P. Hammerstein (Ed.), *Genetic and cultural evolution of cooperation* (pp. 445–468). Cambridge, MA: MIT.

Hillis, K. (1999). *Digital sensations: Space, identity, and embodiment in virtual reality* (1st ed.). Minneapolis, MN: University of Minnesota Press.

Hofstadter, D., & Dennett, D. (Eds.). (1981). *The mind's I: Fantasies and reflections on self and soul*. New York, NY: Basic Books.

Holland, J. H. (1998). *Adaptation in natural and artificial systems: An introductory analysis with applications to biology, control, and artificial intelligence*. Cambridge, MA: MIT.

Honolulu Community College (2007, October). *AEC lunar habitat design team project report*.

Howe, A., & Sherwood, B. (Eds.). (2009). *Out of this world: The new field of space architecture*. Reston, VA: AIAA (American Institute of Aeronautics & Astronautics).

Hooks, B. (1991). Homeplace: A site of resistance. In *Yearning: Race, gender and cultural politics* (pp. 41–49). Toronto: Between the Lines.

Hubbard, S. G. (2005). Humans and robots: Hand in grip. *Acta Astronautica, 57*(2–8), 649–660. doi:10.1016/j.actaastro.2005.03.054.

Hughes, J. (2004a). *Citizen cyborg: Why democratic societies must respond to the redesigned human of the future*. New York, NY: Basic Books.

Hughes, T. P. (2004b). *Human-built world: How to think about technology and culture*. Chicago, IL: University of Chicago Press.

Ihle, E. C., Ritsher, J. B., & Kanas, N. (2006). Positive psychological outcomes of spaceflight: An empirical study. *Aviation, Space, and Environmental Medicine, 77*(2), 93–101.

Imhof, B. (2003). *Musings towards a new genre in space architecture: Transcripts of an architectural journey*. Vienna: Kunst Bundeskanzleramt.

International Space University Student Report (1999). *Out of the cradle: An international strategy for human exploration away from Earth*. Strasbourg, France.

Jain, S. (2006). The prosthetic imagination. In M. Smith & J. Morra (Eds.), *The Prosthetic Impulse: From a Posthuman Present to a Biocultural Future*. Cambridge, Massachusetts: MIT.

Johnson, R., & Holbrow, C. (Eds.). (1977). *Space settlements: A design study*. Washington, DC: NASA Scientific & Technical Information Office.

Johnson, J. C., & Finney, B. R. (1986). Structural approaches to the study of groups in space: A look at two analogs. *Journal of Social Behavior & Personality, 1*(3), 325–347.

Joy, B. (2000). Why the future doesn't need us. *Nanoethics–the ethical and social implications of nanotechnology*, 17–39.

Kanas, N. A., Salnitskiy, V. P., Ritsher, J. B., Gushin, V. I., Weiss, D. S., Saylor, S. A., & Marmar, C. R. (2006). Human interactions in space: ISS vs. Shuttle/Mir. *Acta Astronautica, 59*(1–5), 413–419. doi:10.1016/j.actaastro.2006.02.007.

Kanas, N., & Manzey, D. (2008). *Space psychology and psychiatry*. El Segundo, CA: Microcosm Press/Springer.

Kantermann, T., & Roenneberg, T. (2009). Is light-at-night a health risk factor or a health risk predictor? *Chronobiology International, 26*(6), 1069–1074.

Kaplan, S. (1992). Perception and landscape: Conceptions and misconceptions. In J. Nasar (Ed.), *Environmental aesthetics: Theory, research, and application* (pp. 45–55). Cambridge: Cambridge University Press.

Kaplan, S. (1988). Perception and landscape: Conceptions and misconceptions. In J. Nasar (Ed.), Environmental aesthetics: Theory, research, and application (pp. 45–55). Cambridge: Cambridge University Press.

Kelly, K. (1994). *Out of control: The new biology of machines, social systems, and the economic world*. New York, NY: Basic Books.

Kelly, K. (2010). *What technology wants*. New York, NY: Viking Adult.

Kennedy, K. J. (2002, October). The vernacular of space architecture (AIAA 2002-6102). *1st Space Architecture Symposium (SAS 2002)* (pp. 10–11). Houston, Texas.

Kilgore, D. W. D. (2003). *Astrofuturism: Science, race, and visions of utopia in space*. Philadelphia, PA: University of Pennsylvania Press.

Kirsch, S. (1995). The incredible shrinking world? Technology and the production of space. *Environment and Planning D, 13*, 529–529.

Kitmacher, G. H. (October 2002). Design of the space station habitable modules. In *53rd International Astronautical Congress, in Houston, TX* (pp. 10–19).
Kloog, I., Stevens, R., Haim, A., & Portnov, B. (2010). Nighttime light level co-distributes with breast cancer incidence worldwide. *Cancer Causes & Control, 21*(12), 2059–2068.
Kopec, D. (2009). *Health, sustainability, and the built environment*. New York, NY: Fairchild Books.
Kozicki, J., & Kozicka, J. (2011). Human friendly architectural design for a small Martian base. *Advances in Space Research, 48*(12), 1997–2004.
Knight, D. C. (1977). *Colonies in orbit: The coming age of human settlements in space*. New York, NY: Morrow & Co.
Kramer, W. (2012). *Bioethical considerations and property rights issues associated with the discovery of extraterrestrial biological entities: Implications for political policy in the context of futures studies*. Doctoral dissertation, University of Hawai'i at Manoa.
Krone, B. (Ed.). (2006). *Beyond earth: The future of humans in space*. Ontario, Canada: Apogee Books.
Kropotkin, P. (1995). *Evolution and environment*. Montreal, Canada: Black Rose Books.
Kropotkin, P. (2006). *Mutual aid: A factor of evolution*. Mineola, NY: Dover.
Kunstler, J. H. (1993). *The geography of nowhere: The rise and decline of America's man-made landscape*. New York, NY: Touchstone/Simon & Schuster.
Kurzweil, R. (2005). *The singularity is near: When humans transcend biology*. New York, NY: Penguin Books.
Landis, G. A. (2003). Robots and humans: Synergy in planetary exploration. *AIP Conference Proceedings, 654*(1), 853–860.
Lanier, J. (2010). *You are not a gadget*. New York, NY: Random House.
Launius, R. D., & McCurdy, H. E. (2001). *Imagining space: Achievements, predictions, possibilities: 1950–2050*. San Francisco, CA: Chronicle Books.
Lee, B. (2006). Empathy, androids and "authentic experience". *Connection Science, 18*(4), 419–428.
LeFebvre, H. (1991). *The production of space*. Oxford: Blackwell.
Levy, D. (2009a). The ethical treatment of artificially conscious robots. *International Journal of Social Robotics, 1*(3), 209–216.
Levy, D. (2009). *Love and sex with robots*. HarperCollins e-books.
Lockard, E. S. (2005). *Design considerations for long-term habitation in space: The role of aesthetics and human cognition*. Presented at the 10th Annual International Symposium on Space Exploration. International Space University, Strasbourg, France.
Lockard, E. S. (2006, September). Habitation in space: The relationship between aesthetics and dwelling. AIAA 2006-7331.
Lovelock, J. (2000). *Gaia: A new look at life on earth*. Oxford, England: Oxford University Press.
MacDorman, K. F. (2006, July). Subjective ratings of robot video clips for human likeness, familiarity, and eeriness: An exploration of the uncanny valley. In *ICCS/CogSci-2006 long symposium: Toward social mechanisms of android science* (pp. 26–29).
Mackenzie, B. (2007). Issues of Raising the First Children in Space. AIAA 2007–6077.
Malabou, C. (2008). *What should we do with our brain?* New York, NY: Fordham University Press.
Mander, J. (1991). *In the absence of the sacred: The failure of technology and the survival of the Indian nations*. San Francisco, CA: Sierra Book Club.
Margulis, L. (1998). *Symbiotic planet: A new look at evolution*. New York, NY: Basic Books.
Margulis, L., & Sagan, D. (1997a). *Slanted Truths: Essays on gaia, symbiosis and evolution*. New York, NY: Springer.
Margulis, L., & Sagan, D. (1997b). *Microcosmos: Four billion years of microbial evolution* (1st ed.). Berkeley, CA: University of California Press.
Margulis, L., & Sagan, D. (2003). *Acquiring genomes: The theory of the origins of the species*. New York, NY: Basic Books.

Marston, S. A., & Smith, N. (2001). States, scales and households: Limits to scalar thinking? A response to Brenner. *Progress in Human Geography, 25*(4), 615–619.
Martino, J. P. (1980). Technological forecasting—An overview. *Management Science, 26*(1), 28–33.
Maruyama, M. (1963). The second cybernetics: Deviation-amplifying mutual causal processes. *American Scientist, 51*(2), 164–179.
Maruyama, M. (1975). *Cultures beyond the earth*. New York, NY: Vintage Books.
Maruyama, M. (1978a). Symbiotization of cultural heterogeneity: Scientific, epistemological, and aesthetic bases. In M. Maruyama & A. Harkins (Eds.), *Cultures of the future*. Berlin: De Gruyter Mouton.
Maruyama, M. (1978b). Toward human futuristics. In M. Maruyama & A. Harkin (Eds.), *Cultures of the future* (pp. 33–60). Boston, MA: DeGruyter.
Maruyama, M. (1984). Aesthetics and the environment in outer space, subterranean and underwater communities. *Futures, 16*(2), 148–162.
Maruyama, M. (1994). *Mindscapes in management: Use of individual differences in multicultural management*. Aldershot, England: Dartmouth.
Masini, E. (Ed.). (1983). *Visions of desirable societies*. Oxford, England: Pergamon.
Massey, D. (1994). *A place called home? Space, place and gender (pp. 157-173)*. Minneapolis, MN: University of Minnesota Press.
McCarthy, J. J., & Wright, P. (2007). *Technology as experience*. Cambridge, MA: MIT.
McDonough, W., & Braungart, M. (2002). *Cradle to cradle: Remaking the way we make things*. New York, NY: North Point.
McKenzie, R. D. (1925). *The ecological approach to the study of the human community*. Chicago, IL: University of Chicago Press.
McKibben, B. (2003). *Enough: Staying human in an engineered age*. New York, NY: Holt & Co.
McLuhan, M., & Lapham, L. H. (1964). *Understanding media: The extensions of man*. New York, NY: McGraw-Hill Book Co.
Merleau-Ponty, M. (1962). *Phenomenology of perception*. London: Routledge.
Miedaner, T. (1981). The soul of the Mark III Beast. In D. Hofstadter & D. Dennett (Eds.), *The mind's eye: Fantasies and reflections on self and soul* (pp. 109–113). Toronto, Canada: Bantam Books.
Miller, P. E., & Swinehart, K. D. (2009). *Technological forecasting: A strategic imperative* (pp. 1–4). Johnson City, TN: Department of Management and Marketing, College of Business and Technology, East Tennessee State University.
Monroe, K. M., & Etow, A. (2008). Altruism and cooperation. In B. Sullivan, M. Snyder, & J. Sullivan (Eds.), *Cooperation: The political psychology of effective human interaction* (pp. 89–104). Hoboken, NJ: Blackwell.
Montabone, L. (2008). Future climate changes on Mars: Science fiction or possible reality? In *Planet Mars Research Focus*. Haupaugge, NY: Nova Science Publishers.
Moore, W. (1972). *Technology and social change*. Chicago, IL: Quadrangle Books.
Moran, E. F. (2000). *Human adaptability: An introduction to ecological anthropology*. Boulder, CO: Westview.
Moravec, H. (1988). *Mind children: The future of robot and human intelligence*. Cambridge, MA: Harvard University Press.
Morgan, C. (1980). *Future Man*. Devon, England: Irvington.
Mori, M. (1970). The uncanny valley. *Energy, 7*(4), 33–35.
Morris, B. (2002). Kropotkin's ethical naturalism. *Democracy & Nature, 8*(3), 423–437.
Moser, M. A., & MacLeod, D. (Eds.). (1996). *Immersed in technology: Art and virtual environments*. Cambridge, MA: MIT.
Murphy, M. (2003). *The future of the body: Explorations into the further evolution of human nature*. New York, NY: Penguin Publishers.
NASA Report. *Living in space: Considerations for planning human habitats beyond earth*.
NASA STD-3000 Volumes I–III. (2005). Retrieved from www.standards.nasa.gov/documents

Negroponte, N. (1996). *Being digital.* New York, NY: Vintage Books.
O'Connor, K. (2008). Cooperation in negotiation and conflict resolution. In B. Sullivan, M. Snyder, & J. Sullivan (Eds.), *Cooperation: The political psychology of effective human interaction* (pp. 275–290). Hoboken, NJ: Blackwell.
O'Neill, G. K. (2000). *The high frontier: Human colonies in space.* Ontario, Canada: Apogee Books.
Ondrejka, C. (2007). Collapsing geography (second life, innovation, and the future of national power). *Innovations, 2*(3), 27–54.
Oye, K. (Ed.). (1986a). *Cooperation under anarchy.* Princeton, NJ: Princeton University Press.
Oye, K. (1986b). Explaining cooperation under anarchy: Hypotheses and strategies. In K. Oye (Ed.), *Cooperation under anarchy* (pp. 1–24). Princeton, NJ: Princeton University Press.
Pass, J. (2006a). Astrosociology as the missing perspective. *Astropolitics, 4*(1), 85–99.
Pass, J. (2006b). The astrosociology of space colonies; or the social construction of societies in space. *AIP Conference Proceedings, 813*(1), 1153–1161.
Pepperell, R. (2003). *The post-human condition.* Portland, OR: Intellect Books.
Piliavin, J. (2008). Long-term benefits of habitual helping: Doing well by doing good. In B. Sullivan, M. Snyder, & J. Sullivan (Eds.), *Cooperation: The political psychology of effective human interaction* (pp. 241–258). Hoboken, NJ: Blackwell.
Pink, D. H. (2006). *A whole new mind: Moving from the information age to the conceptual age.* New York, NY: Riverhead Books.
Pinni, M. (2003). *Interior design for a space module.* Reston, VA: American Institute of Aeronautics and Astronautics.
Pólit-Casillas, R. (2010). *Applied space architecture.* Reston, VA: American Institute of Aeronautics and Astronautics.
Poynter, J. (2006). *The human experiment: Two years and twenty minutes inside biosphere 2.* New York, NY: Thunder Mouth.
Prairie View A&M University, Department of Architecture, NASA/USRA Advanced Design Program. (1991). *Mars habitat (design proposal).* Prairie View, TX: Author.
Pratt, M. (1992). Science, planetary consciousness, interiors. In *Imperial eyes: Travel writing and transculturation* (pp. 15–37). USA: Taylor & Francis.
Preiser, W. (1991). Environmental design cybernetics: A relativistic conceptual framework for the design of space stations and settlements. In A. Harrison, Y. Clearwater, & C. McKay (Eds.), *From Antarctica to outer space* (pp. 147–160). New York, NY: Springer.
Radcliff, B. (2008). The politics of human happiness. In B. Sullivan, M. Snyder, & J. Sullivan (Eds.), *Cooperation: The political psychology of effective human interaction* (pp. 305–322). Hoboken, NJ: Blackwell.
Radin, D. (2006). *Entangled minds: Extrasensory experiences in a quantum reality.* New York, NY: Paraview Pocket Books.
Ramey, C. H. (2005). For the sake of others: The personal ethics of human-android interaction. In *Proceedings of the CogSci 2005 Workshop: Toward Social Mechanisms of Android Science in Stresa, Italy* (pp. 137–148).
Rheingold, H. (1992). *Virtual reality: The revolutionary technology of computer-generated artificial worlds—and how it promises to transform society.* New York, NY: Simon & Schuster.
Richerson, P., Boyd, R., & Henrich, J. (2003). Cultural evolution of human cooperation. In P. Hammerstein (Ed.), *Genetic and cultural evolution of cooperation* (pp. 357–388). Cambridge, MA: MIT.
Ritsher, J. B., Ihle, E. C., & Kanas, N. (2005). Positive psychological effects of space missions. *Acta Astronautica, 57*, 630–633.
Ritsher, J., Kanas, N., Ihle, E., & Saylor, S. (2007). Psychological adaptation and salutogenesis in space: Lessons from a series of studies. *Acta Astronautica, 60*, 336–340.
Robinson, J. (1998). Comunidad Los Horcones: Radical Behaviorism in Mexico. *Compendio de Artículos publicados por la Comunidad Los Horcones*, 143–153.

Sale, K. (2000). *Dwellers in the land: The bioregional vision*. Athens, GA: University of Georgia Press.

Sarup, M. (1994). Home and identity. In G. Robertson et al. (Eds.), *Travelers' tales: Narratives of home and displacement* (pp. 93–113). New York, NY: Routledge.

Sasakawa International Center for Space Architecture. (1988). Living in space: Considerations for planning human habitats beyond earth. *Special Information Topic Issue* (Vol. 1, No. 9), Oct–Dec 1988. Houston, TX: University of Houston College of Architecture.

Schutkowski, H. (2006). *Human ecology: Biocultural adaptations in human communities*. Berlin: Springer.

Schwartz, D. (1986). The colonizing experience: A cross-cultural perspective. In B. Finney & E. Jones (Eds.), *Interstellar migration and the human experience* (pp. 234–247). Berkeley/Los Angeles, CA: University of California Press.

Seedhouse, E. (2009). *Martian outpost: The challenges of establishing a human settlement on Mars*. Chichester, UK: Praxis.

Seguin, A. (2005). Engaging space: extraterrestrial architecture and the human psyche. *Acta Astronautica, 56*(9), 980–995.

Seidel, A. (2008). *Inhuman thoughts: Philosophical explorations of posthumanity*. Lantham, MD: Lexington Books.

Senosiain, J. (2003). *Bio-architecture*. Oxford, England: Architectural Press.

Severinson-Eklundh, K., Green, A., & Hüttenrauch, H. (2003). Social and collaborative aspects of interaction with a service robot. *Robotics and Autonomous Systems, 42*(3–4), 223–234.

Shaw-Garlock, G. (2009). Looking forward to sociable robots. *International Journal of Social Robotics, 1*(3), 249–260.

Shea, C., Leveton, L., Slack, K. J., Keeton, K. E., & Palinkas, L. A. (2009). *Antarctica meta-analysis: Psychosocial factors related to long duration isolation and confinement*. Final Report submitted to NASA Behavioral Health and Performance Element.

Sheldrake, R. (1995). *A new science of life: The hypothesis of morphic resonance*. Rochester, VT: Park Street.

Sims, J. H., & Duane, D. B. (Eds.). (1974). *Human behavior and the environment: Interactions between man and his physical world*. Chicago, IL: Maaroufa.

Smith, J. M. (1958). *The theory of evolution*. Baltimore, MD: Pelican Books.

Smith, M. (Ed.). (2005). *Stelarc: The monograph*. Cambridge, MA: MIT.

Smith, M., & Morra, J. (Eds.). (2006). *The prosthetic impulse: From a posthuman present to a biocultural future*. Cambridge, MA: MIT.

Space Science Board, National Academy of Sciences. (1972). *Human factors in long-duration spaceflight*. Washington, DC: National Research Council.

Stallabrass, J. (1995). Empowering technology: The exploration of cyberspace. *New Left Review, 211*, 3–32.

Steiner, D. F. R. (2002). *Human ecology: Following nature's lead*. Washington, DC: Island.

Stelarc. (2000). From psycho-body to cyber-systems: Images as posthuman entities. In D. Bell & B. Kennedy (Eds.), *The cybercultures reader* (pp. 560–576). London: Routledge.

Stevens, R. (2009). Light-at-night, circadian disruption and breast cancer: assessment of existing evidence. *International Journal of Epidemiology, 38*(4), 963–970.

Stock, G. (1993). *Metaman: The merging of humans and machines into a global superorganism*. New York, NY: Simon & Schuster.

Stock, G. (2002). *Redesigning humans: Choosing our genes, changing our future*. New York, NY: Houghton Mifflin Books.

Stone, A. R. (1996). *The war of desire and technology at the close of the mechanical age*. Cambridge, MA: MIT.

Stuster, J. (1996). *Bold endeavors: Lessons from polar and space exploration*. Annapolis, MD: Naval Institute Press.

Suedfeld, P. (1991). Groups in isolation and confinement: Environments and experiences. In A. Harrison, Y. Clearwater, & C. McKay (Eds.), *From Antarctica to outer space: Life in isolation and confinement* (pp. 135–146). New York, NY: Springer.
Suedfeld, P. (1997). Homo invictus: The indomitable species. *Canadian Psychology/Psychologie canadienne, 38*(3), 164–173.
Suedfeld, P. (2005). Invulnerability, coping, salutogenesis, integration: Four phases of space psychology. *Aviation, Space, and Environmental Medicine, 76*(Suppl. 1), B61–B66.
Suedfeld, P. (2010). Mars: Anticipating the next great exploration. Psychology, culture and camaraderie. *Journal of Cosmology, 12*, 3723–3740.
Suedfeld, P., Brcic, J., & Legkaia, K. (2009). Coping with the problems of space flight: Reports from astronauts and cosmonauts. *Acta Astronautica, 65*(3–4), 312–324.
Suedfeld, P., & Steel, G. D. (2000). The environmental psychology of capsule habitats. *Annual Review of Psychology, 51*(1), 227–253.
Sullivan, B. A., Snyder, M., & Sullivan, J. L. (Eds.). (2008a). *Cooperation: The political psychology of effective human interaction*. Hoboken, NJ: Blackwell.
Sullivan, B. A., Snyder, M., & Sullivan, J. (2008b). The centrality of cooperation in the functioning of individuals and groups. In B. Sullivan, M. Snyder, & J. Sullivan (Eds.), *Cooperation: The political psychology of effective human interaction* (pp. 1–16). Hoboken, NJ: Blackwell.
Sullivan, B. A., Snyder, M., Sullivan, J., & Chapp, C. (2008c). The political psychology of cooperation: Synthesis and prospects. In B. Sullivan, M. Snyder, & J. Sullivan (Eds.), *Cooperation: The political psychology of effective human interaction* (pp. 343–358). Hoboken, NJ: Blackwell.
Swift, D. W. (September 2006). *The Voyager spacecraft project—Reasons to remember the grand tour*. AIAA Technical Conference 2006-7463. Reston, VA: American Institute of Aeronautics and Astronautics.
Swyngedouw, E. (2004). Scaled geographies: Nature, place and the politics of scale. In E. Sheppard & R. B. McMaster (Eds.), *Scale and geographic inquiry* (pp. 129–153). Oxford, England: Wiley-Blackwell.
Tanner, N. M. (1986). Interstellar migrations: The beginnings of a familiar process in a new context. In B. Finney & E. Jones (Eds.), *Interstellar migration and the human experience* (pp. 220–233). Berkeley/Los Angeles, CA: University of California Press.
Tapscott, D., & Williams, A. D. (2010). *Wikinomics: How mass collaboration changes everything*. New York, NY: The Penguin Group.
Taylor, M. (1976). *Anarchy and cooperation*. Hoboken, NJ: John Wiley & Sons Ltd.
Taylor, M. (1982). *Community, anarchy and liberty*. Cambridge, MA: Cambridge University Press.
Terraforming Mars: Experts debate how, why, and whether. Retrieved from www.space.com
Terraforming Mars, the noble experiment: An interview with Robert Zubrin. Retrieved from www.astrobio.net
The great terraforming debate. Retrieved from www.astrobio.net
The Terraforming Debates. Retrieved from www.space.com
Thieman, W., & Palladino, M. (2007). *Introduction to Biotechnology*. San Francisco, CA: Pearson Education Incorporated.
Tuan, Y. F. (1974). *Topophilia: A study of environmental perception, attitudes, and values*. New York, NY: Columbia University Press.
Tuan, Y. F. (1977). *Space and place: The perspective of experience*. Minneapolis, MN: University of Minnesota Press.
Turkle, S. (2005). *The second self: Computers and the human spirit*. Cambridge, MA: MIT.
Turkle, S. (2007). *Evocative objects: Things we think with*. Cambridge, MA: MIT.
Tyler, T. (2008). The psychology of cooperation. In B. Sullivan, M. Snyder, & J. Sullivan (Eds.), *Cooperation: The political psychology of effective human interaction* (pp. 105–122). Hoboken, NJ: Blackwell.

Ulrich, R. S. (1991). Effects of interior design on wellness: Theory and recent scientific research. *Journal of Health Care Interior Design, 3*(1), 97–109.
University of Puerto Rico, School of Architecture, NASA/USRA Advanced Design Program. *Camelot III: Habitability Criteria.* (Space Research and Design Studio).
Valentine, J. (1986). The origins of evolutionary novelty and galactic colonization. In B. Finney & E. Jones (Eds.), *Interstellar migration and the human experience* (pp. 266–277). Berkeley/Los Angeles, CA: University of California Press.
Van der Ryn, S. (2005). *Design for life*. Salt Lake City, UT: Gibbs Smith.
Vernadsky, V. I. (1996). *The biosphere: Complete annotated edition*. London: Synergetic.
Vogel, S. (2000). *Cats' paws and catapults: Mechanical worlds of nature and people*. New York, NY: W. W. Norton & Co.
Vogler, A., & Jorgensen, J. (2004, May). Windows to the world; doors to space: A reflection on the psychology and anthropology of space architecture. *Space: Science, Technology and the Arts (7th Workshop on Space and the Arts)* (pp. 18–21). Noordwijk, The Netherlands: ESA/ESTEC.
Vogler, A. (2005). The Universal House – An Outlook to Space-Age Housing. In M. Eekhout (Ed.), *Concept House: Towards customized industrial housing* (pp. 77–87). 1st Concept House Symposium, Delft, The Netherlands.
Vogler, A. (2004). Change of Paradigms: Designing habitats v. designing machines. NASA Explorations Systems White paper.
Walters, M. L., Syrdal, D. S., Dautenhahn, K., Te Boekhorst, R., & Koay, K. L. (2008). Avoiding the uncanny valley: Robot appearance, personality and consistency of behavior in an attention-seeking home scenario for a robot companion. *Autonomous Robots, 24*(2), 159–178.
Walton, J. R., & McConocha, D. M. (1996). Relational dimensions of dog ownership. *Psychological Reports, 79*(3), 947–950.
Wapniak, A. A., Dudley-Flores, M., & Gangale, T. (2007). *An astrosociological approach to defining indigenous Martian architecture*. Reston, VA: American Institute of Aeronautics and Astronautics.
Ward, P. (1995). *End of evolution*. London: Phoenix Grants.
Ward, P. (2001). *Future evolution* (1st ed.). New York, NY: Henry Holt.
Ward, P. (2009). What will become of Homo sapiens? *Scientific American, 300*(1), 68–73.
Weiss, A., Igelsböck, J., Wurhofer, D., & Tscheligi, M. (2011). Looking forward to a "robotic society"? *International Journal of Social Robotics, 3*(2), 111–123.
Weiss, A., Wurhofer, D., & Tscheligi, M. (2009). "I love this dog"—Children's emotional attachment to the robotic dog AIBO. *International Journal of Social Robotics, 1*(3), 243–248.
White, F. (1998). *The overview effect: Space exploration and human evolution*. Reston, VA: American Institute of Aeronautics and Astronautics.
Wiener, N. (1950). *Human use of human beings*. Boston, MA: Houghton Mifflin.
Wilson, E. O. (2003). *The future of life*. New York, NY: Vintage Books.
Winner, L. (1989). *Autonomous technology: Technics-out-of-control as a theme in political thought*. Cambridge, MA: MIT.
Winner, L. (1986). *The whale and the reactor: A search for limits in an age of high technology*. Chicago, IL: University of Chicago Press.
Worringer, W. (1997). *Abstraction and empathy: A contribution to the psychology of style*. Chicago, IL: Elephant Paperbacks.
Zerzan, J. (2002). *Running on emptiness: The pathology of civilization*. Los Angeles, CA: Feral House.
Zubrin, R. (1999). *Entering space: creating a spacefaring civilization*. New York, NY: Tarcher/Putnam.
Zubrin, R. (2003). *Mars on earth*. New York, NY: Tarcher/Penguin.
Zubrin, R. (2008). *How to live on Mars: A trusty guidebook to surviving and thriving on the Red Planet*. New York, NY: Three Rivers.
Zubrin, R., & Wagner, R. (1996). *The Case for Mars: The plan to settle the Red Planet and why we must*. New York, NY: Simon and Schuster.

Zukowsky, J. (1999). *Space architecture*. Stuttgart Germany: Edition Axel Menges.
Zukowsky, J. (2001). *2001: Building for space travel*. New York, NY: Harry N Abrams.
Zylinska, J. (2002). *The cyborg experiments: The extensions of the body in the media age*. London, England: Continuum.

Internet Sources

www.asc-csa.gc.ca
www.cnsa.gov.cn
www.esa.int/ESA
www.inspirationmars.org
www.isro.org
www.jaxa.jp
www.mars-one.com
www.marssociety.org
www.memeburn.com/wp-content/uploads/Envisioning emerging technology for 2012
www.nasa.gov/exploration
www.russianspaceweb.com
www.space.com
www.spacearchitect.org
www.spacefuture.com
www.spacehabs.com
www.virgingalactic.com